Wheat science — today and tomorrow

Wheat science – today and tomorrow

Edited by

L.T.EVANS AND W.J.PEACOCK

CAMBRIDGE UNIVERSITY PRESS

Cambridge

London   New York   New Rochelle

Melbourne   Sydney

CAMBRIDGE UNIVERSITY PRESS
Cambridge, New York, Melbourne, Madrid, Cape Town, Singapore, São Paulo, Delhi

Cambridge University Press
The Edinburgh Building, Cambridge CB2 8RU, UK

Published in the United States of America by Cambridge University Press, New York

www.cambridge.org
Information on this title: www.cambridge.org/9780521105934

First published 1981
This digitally printed version 2009

A catalogue record for this publication is available from the British Library

ISBN 978-0-521-23793-2 hardback
ISBN 978-0-521-10593-4 paperback

CONTENTS

CONTRIBUTORS

A.H.D. Brown: CSIRO Division of Plant Industry, Canberra, A.C.T., Australia

J.A. Considine: Horticultural Research Institute, Department of Agriculture, Victoria, Australia

E.S. Dennis: CSIRO Division of Plant Industry, Canberra, A.C.T., Australia

C.M. Donald: Waite Agricultural Research Institute, Glen Osmond, S.A., Australia

C.J. Driscoll: Waite Agricultural Research Institute, Glen Osmond, S.A., Australia

L.T. Evans: CSIRO Division of Plant Industry, Canberra, A.C.T., Australia

R.A. Fischer: CSIRO Division of Plant Industry, Canberra, A.C.T., Australia

O.H. Frankel: CSIRO Division of Plant Industry, Canberra, A.C.T., Australia

W.L. Gerlach: CSIRO Division of Plant Industry, Canberra, A.C.T., Australia

J.R. Harlan: Crop Evolution Laboratory, University of Illinois, Urbana, Illinois, U.S.A.

A.K.M.R. Islam: Waite Agricultural Research Institute, Glen Osmond, S.A., Australia

R.B. Knox: Department of Botany, University of Melbourne, Parkville, Victoria, Australia

J.B. Langridge: CSIRO Division of Plant Industry, Canberra, A.C.T., Australia

D.R. Marshall: CSIRO Division of Plant Industry, Canberra, A.C.T., Australia

J.B. Passioura: CSIRO Division of Plant Industry, Canberra, A.C.T., Australia

W.J. Peacock: CSIRO Division of Plant Industry, Canberra, A.C.T., Australia

R. Riley: Agricultural Research Council, London, England

E.R. Sears: Department of Agronomy, University of Missouri, Columbia, Missouri, U.S.A.

K.W. Shepherd: Waite Agricultural Research Institute, Glen Osmond, S.A., Australia

D.H. Simmonds: CSIRO Wheat Research Unit, Sydney, N.S.W., Australia

I.A. Watson: Institute of Plant Breeding, University of Sydney, Sydney, N.S.W., Australia

M.J.D. White: Australian National University, Canberra, A.C.T., Australia

S.G. Wildman: Department of Botany, University of California, Los Angeles, California, U.S.A.

FOREWORD

Of the quite small number of plants on which mankind depends for its food, wheat occupies by far the greatest area and produces the greatest amount of edible dry matter. Although wheat is primarily a crop of temperate regions, rice being more important as a staple food in the tropics, many countries at low latitudes are currently trying to adapt wheat to their environments.

Given its primacy among crops, both historically and in current production, wheat has been, and continues to be, the subject of a great deal of research, by many disciplines and in many countries. Of all plant species, wheat is probably the best understood. It is not our intention, in this book, to provide a comprehensive review of all that research. Rather, our purpose has been to present a series of perspectives on current international research on wheat, with emphasis on the way forward into the future.

Our choice of topics and authors was also shaped by our wish to associate them with a symposium to celebrate the eightieth birthday of Sir Otto Frankel on November 4, 1980. This was held at the CSIRO Division of Plant Industry in Canberra, of which he was Chief from 1951 to 1962. Many of the contributors to this book are, or have been, Otto's colleagues within that Division, which he built up to become a major research centre for the plant sciences. Colleagues from other parts of CSIRO and from the universities of Sydney, Melbourne and Adelaide have also contributed chapters, as have several from research institutes overseas.

This book represents, therefore, our joint and affectionate tribute to a stimulating and generous colleague and friend, who has devoted much of his life to understanding and advancing the improvement and conservation of domesticated plants for the future benefit of mankind.

L.T. Evans and W.J. Peacock

Sir Otto Frankel

ACKNOWLEDGEMENTS

The Editors wish to acknowledge the help of Ann Stafford in the design of this book, of Kerrie Harte for producing the camera-ready copy, of Raymond Gorringe for meticulous editing, and of many other colleagues who have contributed to its preparation. The permission of authors and publishers to use previously published figures is also gratefully acknowledged.

1 THE EARLY HISTORY OF WHEAT: EARLIEST TRACES TO THE SACK OF ROME

J.R. Harlan

Introduction

Wheat is, today, one of the most important of all cultivated plants with respect to human nutrition. It is the only crop so far reported to produce more than 400 million metric tons in a single year (Food and Agriculture Organization, 1978). Most of the production is consumed directly as flour. Bran by-products are fed to animals, but relatively little whole grain goes into animal feeds. Wheat is also one of the more nutritious of cereals and its contribution to the human diet puts it clearly in the first rank of plants that feed the world. It was not always so. The common hexaploid bread wheat was a relative late-comer among cereals and its domination did not occur until historical times. Perhaps for that reason the history of wheat holds a special interest.

Earliest traces

Archaeological research has slowly been building evidence for a long and inti-mate association between wheat and man. Some of the earliest traces of this association are indicated in Table 1.1. Not surprisingly, the greater the antiquity the more slender and fragile is the evidence. This is to be expected. The paper by Wendorf *et al.* (1979) mentions only barley, but the person who prepared the SEM photographs (Stemler, personal communication) and the Polish archaeo-botanist Krystyna Wasylikowa (personal communication) are both of the opinion that some kind of wheat is also present in the sample. The wheat seems to match reference material of wild einkorn better than anything else. Not only is the date remarkably early, some 16 000–15 000 BC, but the location of the site by the Nile near Aswan seems an unlikely place to find wild wheat. Indeed, the nearest wild barley known today is in the Sinai some 600 km distant and the nearest stands of wild einkorn are on Mt Hermon in Lebanon more than 1000 km away. Contamination of archaeological samples is always a possibility, but modern contamination would appear virtually impossible in this case, unless it took place in the laboratory.

Wild einkorn requires a cooler and moister climate than wild barley or even wild emmer (Harlan and Zohary, 1966). The climate of Egypt at the time Wadi Kubbanya was occupied is reported as being hyperarid (Wendorf *et al.*, 1979).

Table 1.1
Early traces of wheat in archaeological sites

	Approx. dates BC	Location	Wheat finds	References*
Epipalaeolithic				
Wadi Kubbanya	16 300–15 000	Nr Aswan, Egypt	W-EK (?)	Wendorf (1979)
Nahal Oren	14 800–13 000	Nr Haifa, Israel	C-EM	Noy (1973)
Mesolithic				
Franchthi Cave	ca. 10 000	S. Greece	Barley (no wheat)	Hansen (1978)
Mureybit	9 000–8 000	N. Syria	W-EK	Van Zeist (1968)
Tell Abu Hureyra	9 000–8 000	N. Syria	W-EK	Hillman (1975)
Aceramic Neolithic (PPN)				
Çayönü	>7 000	S.E. Turkey	W-EK; C-EK W-EM; C-EM	Van Zeist (1972)
Ali Khosh	>7 000	S.W. Iran	W-EK; C-EK; C-EM	Helbaek (1969)
Jericho	PPNA 7 000–6 000 PPNB 6 500–5 700	Palestine	C-EK; C-EM	Hopf (1969)
Ramad	ca. 7 000	Nr Damascus, Syria	C-EK; C-EM; N-W	Van Zeist (1966)
Nahal Oren	PPN 7 000–6 000	Israel	C-EM	Noy (1973)
Beidha	ca. 6 800	S. Jordan	C-EM	Helbaek (1966)
Jarmo	ca. 6 750	N.E. Iraq	W-EK; C-EK W-EM; C-EM	Helbaek (1960, 1964)
Hacilar	PPN 5 700	S.W. Turkey	C-EM	Helbaek (1970)
Mehrgarh	>6 000	Baluchistan	C-EK; C-EM; N-W	Jarrige (1980)

*Senior author only listed

W-EK = Wild einkorn; W-EM = Wild emmer; N-W = Naked wheat; C-EK = Cultivated einkorn; C-EM = Cultivated emmer

Both the wheat and the barley must have required flooding of the Nile to grow there at all. We are not sure of the temperatures at the time but, since much of Europe was under ice, the climate could well have been cool as well as extremely dry. We are unable to resolve all the questions raised by these finds at the present time, but taken at face value, they suggest that the large-grained annual grasses of the Near East have been useful for man since the Upper Palaeolithic.

The site of Nahal Oren is not much later and also raises some intriguing questions. It is a cave site, and three grains of emmer were found sealed under a rock fall in Kebaran cultural materials (Noy *et al.*, 1973). The grains have the morphology of cultivated emmer rather than wild emmer. The identification was made by Dr Maria Hopf, one of the most distinguished and competent archaeobotanists in the field. The find seemed so unusual that it inspired Dennell (1973) to propose that modern wild emmer is not really the progenitor of cultivated emmer but a derived weed and that the true progenitor had a morphology like domesticated races. It is quite true that in some crops it is extremely difficult if not impossible to distinguish truly wild races from derived weedy or naturalized races. But, emmer is not a good example of such a situation, because modern wild emmer is not weedy and tends to disappear under extreme disturbance and overgrazing (Harlan and Zohary, 1966). Again, our information is too scanty and incomplete to resolve the question at this time.

With the advance of time our evidence becomes more abundant and our interpretations more sure. In Syria, two sites have been excavated showing abundant use of wild einkorn in Mesolithic cultural contexts. Tell Mureybit (Van Zeist and Casparie, 1968) and Tell Abu Hureyra (Hillman, 1975) are both located on the banks of the Euphrates in northern Syria and both extend through the ninth millennium BC (9000–8000 BC). Wild einkorn is found in quantity in both tells. Einkorn does not grow in the region today and the nearest stands are some 150 km away in Turkey. Van Zeist has suggested that the people might have harvested the wild grain in the north and brought it to the site. I am more inclined to believe that the climate some 10–11 thousand years ago was such that wild einkorn would have grown near these sites. This would certainly be true if the Wadi Kubbanya material is verified. At later time ranges at Abu Hureyra, cultivated einkorn and emmer were retrieved from Neolithic cultural contexts. The site is now under the waters of an artificial reservoir.

A few grains of wild barley, wild oats and wild lentils, but not of wheat, were found in Franchthi cave dating to about 10 000 BC (Hansen and Renfrew, 1978). Wild peas and almonds are documented for Mesolithic times as well. It seems clear that archaeologists are bringing forth evidence for the use of wild cereals in Mesolithic times and indications that the practice probably extended well into the Upper Palaeolithic. The dates used here are crude, uncorrected radiocarbon dates as reported and are too young. Probably two thousand years could be added to the older dates and one thousand years to the Aceramic Neolithic to bring them more in line with real years, but the best methods of correcting dates have not reached general consensus.

As we move into Aceramic Neolithic sites, sometimes called Pre-pottery Neolithic (PPN), the evidence becomes still more abundant and interpretations are on more solid ground. Both wild and cultivated einkorn were found at Çayönü, Ali Khosh and Jarmo. Wild emmer is not frequently encountered but is represented at Çayönü, Jarmo, possibly Abu Hureyra and the Chalcolithic– Bronze Age site of Arad (Hopf, 1978). Cultivated emmer, however, is found at all the PPN sites listed here. It appears that emmer had already emerged as the dominant wheat of early Neolithic times.

The Neolithic dispersal

By the time that pottery was in general use in the Near East, a stable food producing economy had evolved. Barley and emmer were the main cereals. Pulses included lentil, pea, chickpea, vetches and, in due time, fava beans. Flax, grape, fig, date, pomegranate, almond, pistachio, olive and a variety of vegetables were exploited. Sheep, goats, cattle and swine were usually integrated with crop growing in some fashion. The system was expansive and spread through the Mediterranean world, across the Balkans, up the Danube and down the Rhine. The spread into Europe is documented by hundreds of sites that have now been studied. The time schedule was roughly as follows: into Greece by 6000 BC; to Spain and the lower Danube by 5000 BC; to the Low Countries a little before 4000 BC; to England and Scandinavia by 3000 BC (Milisauskas, 1978). The Egyptian Fayum Neolithic is dated some 4500 BC, but earlier settlements could well be buried under the silts of the delta. Eastward, wheats have been reported from PPN Mehrgahr in Baluchistan, Pakistan at more than 6000 BC and a flourishing cereal agriculture was in place by 5000 BC (Jarrige and Meadow, 1980). Wheat was grown by Indus cultures by 3000 BC at least, and reached China sometime in the second millennium BC, and certainly before 1300 BC (Ho, 1969).

Emmer was the dominant wheat throughout the Neolithic dispersal across Europe and North Africa. Only occasionally do we find large or reasonably pure samples of einkorn. The carbonized grains of einkorn that are recovered are most often minor components of emmer or barley samples and may represent weedy admixtures of these crops. Not only is wild einkorn weedy, but the cultivated races have also been observed as weeds in Armenia (Tumanjan, 1935) and in the Crimea (Barulina, 1924/25). In any case, the two wheats spread together along with barley throughout agricultural Europe. Elsewhere, they parted company. Einkorn did not reach Egypt nor has it been reported elsewhere in North Africa except Morocco where it was probably introduced from Spain. We had no record of it east of western Iran until the recent report of it in Baluchistan (Jarrige and Meadow, 1980). This find should be carefully verified since it currently represents an anomoly in einkorn distribution. It is not known in Ethiopia or Central Asia.

Although a minor cereal from the beginning, einkorn had a remarkable longevity, being recorded in this century in Morocco, Spain, France, Switzerland, Germany, Italy, Bulgaria, Yugoslavia, Greece, Transcaucasia and Turkey (Percival, 1921), and Sweden could be added to this list (Hjelmqvist, 1963). It is, perhaps, an example of the remarkable tenacity of peasant farmers. It can be grown on poor soils and is not demanding of fertility. It is nutritious, but makes a miserable bread (Le Clerc *et al.*, 1918) and was consumed primarily as porridge. Today, most of the production is fed to livestock.

Evidence for an early wave of emmer dispersal is revealed by recent distributions. The Ethiopian, Yemeni and Indian emmers are closely related. This group has more vascular bundles in the coleoptile than European emmers (Percival, 1927a). A scattering of emmer samples from remote regions of Turkey, Iran and Transcaucasia belong to the Afro-Indian group of emmers (Percival, 1926; Flaksberger, 1928). It is likely that these represent relicts of the initial Neolithic dispersal. The emmers of Ancient Egypt were also part of the same group (Flaksberger, 1928). Emmer is still an important crop in Ethiopia and parts of Peninsular India. About 1910 the average production in Russia was some 10 million bushels, mostly from the Volga regions (Carleton, 1911).

It is easy to understand the preference for emmer over einkorn, but it is not easy to understand the preference for a glume wheat like emmer over a free-threshing naked wheat. Archaeology tells us that naked wheats were available from early Neolithic onwards. A few grains were found in Ramad, Syria at ca. 7000 BC (Van Zeist and Bottema, 1966) and a large sample (nearly 3000 grains) from Knossos was found by Dr Helbaek to consist almost entirely of what he called *Triticum aestivum* (Evans, 1968). The uncorrected date is about 6000 BC and a corrected one would be at least 7000 BC. Naked wheat is reported from PPN in Baluchistan at over 6000 BC (Jarrige and Meadow, 1980) yet emmer remained king of the wheats for several thousand years. Naked wheats have turned up in small quantities in many sites of the European Neolithic and Bronze Ages, but did not dominate wheat growing there until later.

At this point a caveat is in order. One would suppose that the early naked wheats would be some form of *Triticum turgidum*, a tetraploid like emmer. But this species almost never appears in archaeological literature dealing with this time range. When archaeobotanists find naked wheat they have usually called it

T. aestivum. One gets the impression from the reported archaeological record that the free-threshing tetraploid wheats have had no evolution at all and have sprung fully-formed from the forehead of Ceres. *T. turgidum* is not in Helbaek's vocabulary; to him all free-threshing wheats are *T. aestivum.* Hopf uses *T. aestivum* s.l. to suggest a broad interpretation of the species. Jarrige and Meadow give us a choice (*T. durum* or *T. aestivum*). After discussing the matter informally with the major archaeobotanists of the field, Zohary (1973) finally raised the point in print: Where are the naked tetraploid wheats? Van Zeist (1976) eventually gave the answer. After an extensive and detailed study, he concluded that it is not possible to separate hexaploid and tetraploid naked wheats on the basis of carbonized grains, rachis forks or other plant parts preserved in archaeological samples.

Some of the earliest naked wheat grains recovered were shaped like club wheats and were, therefore, referred to the hexaploid species. If the shape was due to the effect of the C gene, this would be correct since it is located on chromosome 2D. But there are compact tetraploid club wheats as well, and Kislev (1977) has challenged the correctness of these early identifications. In fact, the enormous variability of both tetraploid and hexaploid wheats was grossly underestimated in earlier archaeobotanical work. Nor was the distortion of shape during the process of carbonization adequately appreciated. A few of the botanists have studied the problem systematically. Hopf (1955), Stewart and Robertson (1971), Tellez and Ciferri (1954) and our laboratory (Harlan, unpublished) have done experimental charring of seeds. I know from experience that it is possible, under the right conditions, to make the slender seed of einkorn swell to the shape of bread wheat. When dealing with small numbers of poorly preserved materials, extreme caution is indicated.

Durum wheat has occasionally been identified archaeologically. Percival (1927b), who certainly knew wheat, had no difficulty in pronouncing the Jemdet Nasr (Iraq) samples as durum. Kislev (n.d.) identified durum wheat at Tell Keisan, an early Iron Age site near Haifa. On the whole, however, the origin and evolution of the naked tetraploid wheats are obscured and confounded by confusion with naked hexaploid wheat. This is unfortunate and it is to be hoped that the real story will some day be unravelled. In any case, the curious fact remains that people continued to grow emmer even when free-threshing wheats were available. The extra processing required of a glume wheat was evidently worth the effort.

The recent distribution of wheats may provide some perspective to the problem. The wheats of Ethiopia, until recent years, were almost all tetraploid, mostly durum and emmer (Vavilov, 1926). This combination, together with barley, was a basic Neolithic mix that spread widely in those times. We do not know when those crops reached the Ethiopian plateau, but the essential absence of hexaploid wheat may be a clue. In Ethiopia, both naked and glumed wheats are grown in the same villages and often by the same individuals. There are, evidently, advantages to both.

On the other hand, the wheats of China are essentially all hexaploid and there is very little tetraploid wheat in the Orient (Vavilov, 1926; Orlov, 1923). India has (or had) a considerable diversity of both durum and bread wheat (Howard and Howard, 1927). The sphaerococcum wheats are largely endemic to north India and Pakistan. These are reported archaeologically in Baluchistan ca. 4000 BC (Jarrige and Meadow, 1980). The tetraploids and hexaploids seem to have had distinctly different times and routes of dispersal.

The Soviet analyses of diversity among the durums were made early enough that authentic samples could be collected (Orlov, 1923). All of the races described were found in Africa, and there was nothing in Europe or Asia not also in Africa. There was a suite of races endemic to Ethiopia, as one might expect; otherwise greatest diversity was found in Algeria and Egypt. On the whole, the durums and other races of *T. turgidum* are Mediterranean in adaptation and diminish in diversity and performance as they are removed from the region. It is tempting to suggest that all of the early naked wheats reported from near the Mediterranean and the Near East were tetraploids and that *T. aestivum* came later and by a different route.

The world of emmer and barley

The wheat of ancient Egypt was emmer, but this may have been due to prejudice. Herodotus, in the fifth century BC wrote:

"... while other nations live on naked wheat and barley, it is considered in Egypt the greatest shame to live on them; they prepare their bread from ὄλυρα which some call Zeía". (Herodotus, II, 36).

The Egyptians had three words for cereals, $\triangleleft$ $\boxminus$, a reed plume, a loaf of bread and three grains is transliterated as j t and meant barley; Ψ or $\lrcorner$ $\rightleftharpoons$ Ψ or other variants showed an ear of bearded grain together with a foot, a hand, a loaf of bread, etc. was transliterated as b d t and meant emmer; while $\longrightarrow\!\!\boxminus\!\!\longleftarrow$ $\mathfrak{L}$ $\boxminus$, a bolt (lock), quail chick, a loaf of bread and three grains was transliterated as s w t (Gardiner, 1957). Hrozný (1913) and other scholars thought the third name referred to durum. But, durum has not turned up archaeologically in Egypt until after the Greek conquest (Dixon, 1969). The s w t must have referred to something else.

Wooden tomb models give us an insight into the processing of grain in Ancient Egypt. The brewery and the bakery always occupied the same building. Figures are shown (usually male) using mortar and pestle to hull the emmer.

Adjacent groups (usually female) are shown grinding hulled grain into flour. This is then mixed with water and baked. Some of the bread is put into water in jars, inoculated and allowed to ferment in the brewer's section of the building. Presumably sources of sugar were added or some of the grain malted. Additives mentioned in Egyptian texts include figs, dates and grapes and various kinds of yeast cultures were given separate names. Wheat beer, called *bouza*, is made the same way today in Cairo (Lucas, 1962). In the modern recipe, a part of the wheat is malted, crushed and added to the bread in the jars. The loaves are baked only lightly so as not to kill the yeast mixed with the dough. According to Lucas (1962) the alcohol content is rather higher than in traditional barley beers.

Emmer wheat and barley are the predominant cereals of Hittite cuneiform literature and are the only cereals listed in Hittite laws (Chadwick, 1973). According to Hrozný (1913), the ancient Babylonian word for emmer was ZIZ with various modifiers such as ZIZ-GAR, emmer for bread, and ZIZ-KAS, emmer for beer. It was written $\bigtriangledown$ $\bigtriangledown$. There were other words for cereals, but we do not know if the ancients of Mesopotamia distinguished between durum and bread wheat. Kislev (1973) has suggested that the ancient Hebrew word *kussemet* meant emmer, the word *hitta* meant naked wheat, probably durum in Palestine and bread wheat in Mesopotamia, and the word *šifon* referred to einkorn.

Emmer was traditionally de-glumed in a mortar throughout the Mediterranean and Near Eastern lands. The Mycenaean glyph for 'flour' was a figure holding a pestle ⚲ or ⚲ (Chadwick, 1973). Roman bakers were called *pistores* from the verb *pinsere*, 'to pound' (the grain) (Smith *et al.*, 1891). The emmer heritage was strong in Rome even after it was little used. It was a sacred grain and called *semen adoreum* or simply *ador* for short, and this in turn became the word 'glory'. (Pliny XVIII, 3). Pliny (XVIII, 11) cites Verrius as stating "emmer was the only cereal used by the Roman nation for 300 years". A yearly celebration was established by Numa called the Feast of the Ovens in which emmer was roasted for a sacrificial meal (Pliny XVIII, 2). Protocols for sacrificing to the gods using emmer are given by Cato and Pliny. Roman recipes even in imperial times often called for emmer groats.

Another Latin word for emmer was *far*, from which the words *farina* (flour) and *farinaceous* (starchy) were derived. The harvest ceremonies of ancient Rome were called *farnacalia* and the traditional marriage vows were *confarati* in which *far* was ritually consumed by the partners. North of the Alps, the same root word was pronounced *bar* and attached to a different cereal. It meant originally a bearded grain and from it such words as barb, barbe, beard, barber, beer, brew, brewing and brewery were derived. For grain it was first used as an adjective as in 'barlic korn' or 'barlie korn'. It was finally reduced to 'barley'. A 'barn' is a place to store barley and 'Barton' a village where barley grew well (Andrews, 1964).

Despite the availability of naked wheat, whether it was durum or bread wheat, the picture given us in early historic times is one dominated by emmer and barley. The Romans rejected barley for human food at an early date, but the Greeks clung tenaciously to their *maza* porridge. Greek tradition asserted it was the strongest and healthiest of cereals. Athletes and soldiers were trained on barley and gladiators were called *hordearii*, or barleymen. (Pliny XVIII, 14).

The rise of bread wheat

While barley and emmer dominated Europe south of the Alps, the Near East and North Africa, other cereals were developing north of the Alps. Oats, rye and spelt wheat evolved as relatively new crops of the region. Carbonized grains of oats and rye have been found in Neolithic contexts, but the early finds seem to represent weeds in other crops, occurring in small numbers among samples of barley and emmer. It was not until Iron Age (for Central Europe c. eighth century BC) that these cereals became important crops in their own right. The story of spelt has been as perplexing as the case of free-threshing tetraploid wheat. Until recently, it seemed to have no reasonable origin, appearing without antecedents in northern Europe.

McFadden and Sears (1946) synthesized hexaploid spelt from an emmer and *Aegilops squarrosa (Triticum tauschii)*, and Kihara (1944) also demonstrated that this wild species is the donor of the D genome. These results indicate very strongly that spelt should have originated within the natural range of *Ae. squarrosa* (Zohary *et al.*, 1969). The crop has been found growing on a small scale in remote sections of Iran (Kuckuck, 1964), Turkey and Transcaucasia (Dorofejev, 1971). The Asian spelts seem to be rather different from European spelts.

Spelt is not known archaeologically from the Near East or Mediterranean regions. Flaksberger (1930) stated flatly that spelt could not be the progenitor of bread wheat but must be a derivative of it. Only a few years ago, Feldman (1976, p. 126) wrote:

> "The genetic data, which show that the hulled hexaploid wheats are more primitive than the free-threshing forms are not in accord with archaeological chronology. While *aestivum* was abundant in the prehistoric Near East from the sixth millennium onwards there are, so far, no indications of the cultivation of *spelta* or other hulled forms before 2000 BC. Moreover, in central Europe, *spelta* appears about 1000 years later than compact forms of free-threshing wheat."

Kerber and Rowland (1974) produced impressive evidence that the first hexaploid wheats must have been non-free-threshing. They synthesized 15 hexaploid wheats from 9 tetraploids and 7 accessions of *Ae. squarrosa*. All were non-free-threshing regardless of the presence or absence of Q. There is a gene for tenacious glumes, *Tg* on chromosome 2D, and it inhibits the expression of Q on chromosome 5A. Consequently, free-threshing hexaploid wheat must have both tg tg and Q Q and mutation to free-threshing must have taken place at the hexaploid level. The only tetraploid wheat with Q is *Triticum carthlicum*.

Riley (1969) pointed out that the cultivar 'Chinese Spring' has a primitive karyotype because there are no chromosome interchanges between it and either wild emmer or *Ae. squarrosa*. Using it as a tester, he could also show that at least some of the Iranian spelts were also primitive. Two European spelts showed one interchange difference as did several bread wheats from Europe, China, India and U.S.A. His sample of *sphaerococcum* showed a difference of two interchanges with Chinese Spring.

The problem of the history of spelt now seems closer to a solution. Lisitsina (1978) has reported archaeological spelt in the Caucasus dating to the fifth millennium BC and possibly to the sixth. Spelt has been found in Moldavia dating between 4000 and 3000 BC and more has been found in Bulgaria and the Black Sea littoral. Janushevich (1978) suggested two routes of dispersal, one from the traditional nuclear area across Anatolia to Greece then to the Danube and central Europe; the second from Transcaucasia northward across the Caucasus, the lower Don and Dnepr Valleys and then to central Europe. Schultze-Motel and Kruse (1965) plotted the distribution of archaeological sites with spelt known at that date. It was as follows:

Stone Age: 2 south Germany, 3 Poland, 1 Sweden
Bronze Age: 5 Switzerland, 2 south Germany, 1 north Italy, 1 Denmark
Hallstatt: 1 central Germany, 5 north Poland
Iron Age: 1 Alsace, 13 southern England, 1 north Germany, 5 Sweden

The various ages came to different parts of Europe at different times so that a true chronology of events is complicated and not easily constructed. Since this report, spelt has been found in Czechoslovakia of Hallstatt times (Hajnalova, 1978) and other finds are now consistent with the view of Janushevich (1978) that spelt took a different route to Europe than was taken by barley, emmer and einkorn. Archaeology has now confirmed the history of spelt as proposed by McFadden and Sears in 1946.

Spelt became closely associated with some of the Germanic tribes, and although it is a glume wheat and considered primitive, it lingered on in some abundance in central Europe competing well with naked bread wheat. As late as the first decade of this century the area sown to spelt exceeded that of bread wheat in Württenberg, Swabia, Baden and a number of other districts in southern Germany and German-speaking Switzerland (Gradmann, 1909). It had particular qualities that made it popular in its region of adaptation. It was especially winter hardy and could be autumn-sown farther north and at higher altitudes than bread wheat, barley and oats. It was relatively undemanding on thin mountain soils, and in such environments readily outyielded bread wheat. It was disease resistant, and the flour was of excellent quality. Extracting the grain from the glumes became much less of a problem after mills were adapted for the purpose.

The Romans did not know the northern cereals until their conquests of Gaul, England and some of the German tribes. Pliny describes oats as a 'disease' of wheat and barley. To the Romans, it was a weed even though some of the Germans lived on meal made from it. *Siligo* was a Latin name for a kind of wheat, but it was somehow transformed to *secale* and referred to rye.

Of this crop, Pliny wrote (XVIII, 40) ". . . it is a very poor food and only serves to avert starvation . . . it is of a dark sombre colour, and exceptionally heavy. Emmer is mixed in with this to mitigate its bitter taste, and all the same it is very unacceptable to the stomach even so." 'Spelt', on the other hand, was a Celtic word accepted by the Romans. The first occurrence of the word *spelta* appeared in the Edict of Diocletian, 301 AD when the prices of many commodities and services were controlled. By this time, the Romans had learned to appreciate spelt. Some German colonies were established in the Po Valley to grow spelt and supply it to Roman epicures. Van Zeist (1972) reported that the only archaeological spelt so far found in the Netherlands came from ruins of Roman villas.

Our inability to distinguish the naked tetraploid wheats from the hexaploids leaves the history of the true bread wheats somewhat obscure. Naked wheat was present throughout the European Neolithic and the first European culture to be based on naked wheat is said to be the Cortaillod of Switzerland, ca. 3200–2600 BC (Murray, 1970). We do not know if this wheat was tetraploid or hexaploid, but other cultures continued to live largely on barley and emmer into the Iron Age. The reasons for the slow adoption of naked wheat in Europe are obscure.

In Asia, people were less reluctant. The Soviet analyses of diversity, as well as other studies, indicate that the greatest diversity was generated in a belt from Transcaucasia and eastern Turkey across Iran, Afghanistan into the mountains of Pakistan and India (Vavilov, 1926). Naked wheats were grown in Baluchistan before 6000 BC (Jarrige and Meadow, 1980), but these could well be durums. The people of the Altinova plain in south-eastern Turkey had largely converted from emmer to naked wheat by 3000 BC, but again, we do not know if it was tetraploid, hexaploid or both (Van Zeist and Bakker-Heers, 1975). Wheat culture was thriving in the Indus Valley by the third millennium BC (Vishnu-Mittre, 1974). Bread wheat was introduced to China by the second half of the second millennium BC (Ho, 1969). It is possible that Asian tribes accepted bread wheat more readily because of less competition from emmer, rye, oats, and spelt. But it is a curious fact that bread wheat was accepted as a major crop late in the development of European agriculture. The emergence of the Roman empire had a role to play.

Romans began as emmer-eating people, using the grain mostly as porridge (*puls*) and by the end of the Republic had converted to the use of bread wheat. The rise in popularity of raised bread was closely associated with the economy, standard of living, politics and fate of the empire (Jasny, 1944). Roman historians were in agreement that the great strength and vitality of the early Republic was in the basic support of the independent small farmer. In those days even generals laid their own hands to the plow and worked the fields with pride. Pliny (XVIII, 6) cites Cato to the effect that: "The agricultural class produces the bravest men, the most gallant soldiers and the citizens least given to evil designs." Roman family names often reflected crops which some patriarch was particularly adept at raising: Cicero (chickpea), Lentulus (lentil), and Fabius (fava bean) (Pliny XVIII, 3).

But, as the Republic flourished, as new lands were acquired and more tribes conquered, the traditional rustic life began to change. The city grew enormously; small farms were consolidated into estates and eventually into vast plantations with thousands of slaves, serfs or tenants to work the land. The small holder tended to become economically inviable and farms were abandoned or swallowed up by the rich landowners. Rome could no longer be supplied by local produce. The provisioning of Rome became a major problem, a major bureaucratic function, and an opportunity for political exploitation and power.

In the fourth century BC, Alexander took Egypt and the Ptolemaic dynasty was established after his death in 323 BC. So far as we know, the only wheat grown in Egypt at the time was emmer, but the Ptolemys saw excellent opportunities for foreign trade in bread wheat. The growing Roman market could help the balance of payments. By the first century BC, the flow of grain from North Africa to Rome had become very substantial. The prices fluctuated wildly and Rome was vulnerable, having become dependent on foreign grain suppliers. Extraordinary powers were granted to Scaurus in 104 BC and to Pompey in 57 BC in order to save Rome from starvation. (Stevenson, 1934). After food riots in 62 BC, Cato distributed grain at half price. Clodius did him one better and distributed free grain in 59 BC. The welfare dole had been established and could only be repealed at great political risk.

In the crisis of 22 BC when Rome was again threatened by famine, a mob shut up the Senate in the *Curie* and threatened to burn the building down over their heads if they would not agree to put Augustus in charge of the situation. With the consent of the Senate and the support of the people, Augustus was offered both a dictatorship and the office of *cura annonae*. Consistent with his political stance, Augustus refused the dictatorship, but accepted the office which was charged with the provisioning of Rome. He quickly eased the crisis at his own expense and thereby secured powerful support for the rest of his career. A board of ex-praetors was appointed to supervise the distribution of free wheat. The number of recipients had been fixed by Caesar at 150 000, but somehow managed to rise to 320 000 by 5 BC (Stevenson, 1934). The expense was enormous and absorbed about half the revenues derived from Pompey's conquests.

While Augustus was a master politician, a capable administrator and a general adequate for the defeat of Antony, his power came from control of the legions and of the Roman food supply. The flow of wheat from North Africa to Rome was an obvious object for political ambition. About 9 AD, still under Augustus, the office of *Praefectus annonae* was established and "conferred on C. Turrianus, a man of equestrian rank, who held it until the reign of Claudius. From this time onwards the emperor made himself responsible for the feeding of the city, and his equestrian prefect became the head of a large department with representatives in the ports and in the provinces" (Stevenson, 1934, p. 203). And so, Augustus, born Octavius, adopted son of Caesar, has given his name to the eighth month of our calendar. It should be remembered that wheat power helped keep him in office through a long career.

Archaeological research has located an office building in Rome where the *cura annonae* was quartered. The names of a number of Praefects are known as well as those of officers stationed at Ostia, Alexandria and other ports. The department served the empire for over 300 years (Pavis D'Escurac, 1976). The size of the wheat shipments to Rome was considerable. The figure of 20 million *modii*, or about 5 million bushels of wheat per year has been given for the export of Egypt to Rome in the first century AD (Stevens, 1941), and under Justinian, some 7 million bushels was also sent to Constantinople. According to Josephus, the Egyptian wheat fed Rome for four months of the year and North Africa (Tunisia, Algeria) supplied the grain for eight months. Substantial quantities were also obtained from Sicily, Sardinia, Gaul and the Po Valley. Pavis D'Escurac (1976), however, recommends caution in accepting the reported figures.

Early Republican Rome had no bakers according to Pliny (XVIII, 28); bread was prepared in the home by servants or the housewife. But, as the city increased in size and standards of living rose, baked goods came into demand. The *pistores*, as we have seen, got their name because they used to dehull emmer with a mortar and pestle. As baked goods became a necessity rather than a luxury, a powerful bakers guild was organized, the *collegium pistorum*. The guild was closely connected to the *cura annonae* and its members enjoyed a number of privileges. Some emperors found it expedient to issue not only free grain as a dole, but bread as well. According to one recipe left to us, some Roman bakers reversed the Egyptian procedure of making bread to brew beer and used beer yeast to raise bread (Pliny XVIII, 12).

And so the demand for bread wheat grew, Rome was converted to the use of raised breads and all the wheat-growing lands of the Mediterranean basin were converted to bread wheat in order to supply Rome. Roman taste and custom expanded with the empire. But the conversion was only partial. The Gauls accepted the bread and the vine readily; the Germanic tribes did not. They preferred their beer, their porridge and their dark breads. Did the sack of Rome stabilize this dichotomy? Did the disintegration of the empire result in intensification of tribalism? Did the expansion of white bread fail because Rome failed? The ethnic differences have persisted into modern times when soldiers campaigning in France and Germany noted:

Soldaten trost

Nein! hier hat es keine Not:
Schwartze Mädchen, weisses Brot!
Morgen in ein ander Städtchen:
Schwartzes Brot und weisse Mädchen!
                    (Gradmann, 1909)

References
Andrews, A.C. (1964) The genetic origin of spelt and related wheats. *Züchter* **34**, 17–22.
Barulina, E.I. (1924/25) *Triticum monococcum* as an admixture to cereal crops in the Crimea. *Bull. Appl. Bot. Plant Breed.* (Leningrad) **14**, 136–139.
Carleton, M.A. (1911) Winter emmer. *U.S.D.A. Farmers' Bulletin No. 466*. Washington. p. 24.
Chadwick, J. (1973) Documents in Mycenaean Greek (2nd ed.). The University Press, Cambridge.
Dennell, R.W. (1973) The phylogenesis of *Triticum dicoccum*: a reconsideration. *Econ. Bot.* **27**, 329–331.

Dixon, D.H. (1969) A note on cereals in ancient Egypt. In 'The domestication and exploitation of plants and animals'. (Eds. P.J. Ucko and G.W. Dimbleby.) G. Duckworth, London. pp. 131–142.

Dorofejev, W.F. (1971) Die Weizen Transkaukasiens und ihre Bedeutung in die Evolution der Gattung *Triticum* L. 3. Die Spelzweizen Transkaukasiens. (*T. macha* Dek. et. Men, *T. spelta* L. ssp. *spelta* Dorof. ssp. *kuckuckianum* Gökg. und ssp. *vavilovii* Sears). *Z. Pflanzenzücht.* **66**, 335–360.

Evans, J.D. (1968) Knossos neolithic, Part II. Stone axes and maceheads: materials. *British School of Archaeology, Athens* **63**, 239–276.

Feldman, M. (1976) Wheats. In 'Evolution of crop plants'. (Ed. N.W. Simmonds.) Longman, London. pp. 120–128.

Flaksberger, C. (1928) The emmers (*Triticum dicoccum* Schrank) of ancient Egypt and modern times. *Bull. Appl. Bot. Genet. Plant Breed.* **19** (Part 1), 495–518. Russian and English.

Flaksberger, C. (1930) Ursprungszentrum und geographische Verbreitung des Spelzes (*Triticum spelta* L.). *Angew. Bot.* **12**, 86–99.

Food and Agriculture Organization (1978) Production yearbook. FAO, Rome.

Gardiner, A. (1957) Egyptian grammar (3rd ed.) Griffith Institute, Ashmolean Museum, Oxford.

Gradmann, R. (1909) Der Getreidebau im deutschen und römischen Altertum. Jena, Hermann Costenoble. p. 111.

Hajnalova, E. (1978) Funde von *Triticum*-Resten aus einer hallstattzeitlichen Getreidespeichergrube in Bratislava-Devín/CSSR. *Ber. Dtsch. Bot. Ges.* **91**, 85–96.

Hansen, J. and Renfrew, J.M. (1978) Palaeolithic–Neolithic seed remains at Franchthi Cave, Greece. *Nature* **271**, 349–352.

Harlan, J.R. and Zohary, D. (1966) Distribution of wild wheats and barley. *Science* **153**, 1074–1080.

Helbaek, H. (1960) The paleobotany of the near east and Europe. In 'Prehistoric investigations in Iraqi Kurdestan'. Studies in Ancient Oriental civilization No. 31. 184 pp. 29 plates, illus. (Eds. R.J. Braidwood and B. Howe.) Univ. Chicago Press, Chicago.

Helbaek, H. (1964) Archaeological evidence for genetical changes in wheat and barley. Int. Bot. Congr. (Abstr.). Edinburg, August, 1964.

Helbaek, H. (1966) Pre-pottery farming at Beidha: a preliminary report. In 'Five seasons at the pre-pottery village of Beidha in Jordan'. (Ed. D.D. Kirkbride.) pp. 61–66. Palestine Explor. Quart. 1966, 8–72.

Helbaek, H. (1969) Plant collecting, dry farming and irrigation in prehistoric Deh Luran. In 'Prehistory and human ecology of the Deh Luran plain'. (Ed. F. Hole.) Mem. Mus. Anthrop. Univ. Michigan No. 1. Ann Arbor. pp. 383–426.

Helbaek, H. (1970) The plant husbandry of Hacilar. In 'Excavations at Hacilar'. (Ed. J. Mellaart.) British Inst. Archaeol. Ankara. Occas. Pap No. 9. Edinburgh Univ. Press. pp. 189–244.

Herodotus (1934) History. Transl. by George Rawlinson. Tudor Publishing Co., New York.

Hillman, G. (1975) The plant remains from Tell Abu Hureyra: A preliminary report. *Proc. Prehist. Soc.* **41**, 70–73.

Hjelmqvist, H. (1963) Zur Geschichte des Einkorns and des Emmers in Schweden. *Bot. Not.* **11**, 487–497.

Ho, P.T. (1969) The loess and the origin of Chinese agriculture. *Am. Historical Rev.* **75**(1), 1–36.

Hopf, M. (1955) Formveränderungen von Getreidekornern beim Verkohlen. *Ber. Dtsch Bot. Ges.* **68**, 191–193.

Hopf, M. (1969) Plant remains and early farming in Jericho . In 'The domestication and exploitation of plants and animals'. (Eds. P.J. Ucko and G.W. Dimbleby.)   Chicago, Aldine Publishing Company. pp. 355–359.

Hopf, M. (1978) Plant remains, Strata V–I. In 'Early Arad I. The chalcolithic settlement and early bronze city'. (Ed. R. Amiran.) Israel Explor. Soc., Jerusalem. pp. 64–82.

Howard, A. and Howard, G.L.C. (1927) Improvement of Indian Wheat. Pusa Agr. Res. Inst. Bull. No. 171. Pusa (India). p. 26.

Hrozny, F. (1913) Das Getreide im alten Babylonien. *Sitzungsber. Kais. Akad. Wiss. Wien* **173**, 1–218.

Janushevich, Z.V. (1978) Prehistoric food plants in south-west of the Soviet Union. *Ber. Dtsch. Bot. Ges.* **91**, 59–66.

Jarrige, J.F. and Meadow, R.H. (1980) The antecedents of civilization in the Indus Valley. *Sci. Am.* **243**(2), 122–125, 128–130, 132–133.

Jasny, N. (1944) The wheats of classical antiquity. *Stud. Hist. Polit. Sci. Series 62, No. 3.* Johns Hopkins Press, Baltimore. p. 176.

Kerber, E.R. and Rowland, G.G. (1974) Origin of the free-threshing character in hexaploid wheat. *Can. J. Genet. Cytol.* **16**, 145–154.

Kihara, H. (1944) Die Entdeckung des DD-analysators beim Weizen. *Agric. Hortic. Japan* **19**, 889–890.

Kislev, M. (1973) Hitta and Kussemet, Notes on their Interpretation, Lešonenu Vol XXXVII No. 2–3. pp. 83–95.

Kislev, M. (1977) Is it possible to identify local wheats of the Near East by their compactness? International Work Group for Paleoethnobotany IV. Symposium IAP/IWGP.

Kislev, M. (n.d.) Contenu d'un silo à blé d'époque israélite ancienne. Typed Ms. Universite Bar-Ilan, Ramat-Gan.

Kuckuck, H. (1964) Experimentelle Untersuchungen zur Entstehung der Kultur-weizen. I. Die Variation des iranischen Spelzweizens und seine genetischen Beziehung zu *Triticum aestivum* ssp. *vulgare* (Vill., Host.) MacKey, spp. *spelta* (L). Thell. und ssp. *macha* (Dek. et Men.) MacKey mit einem Beitrag zur Genetik des Spelta-Komplexes. *Z. Pflanzenzücht.* 51, 97–140.

Le Clerc, J.A., Bailey, L.H. and Wessling, H.L. (1918) Milling and baking tests of einkorn, emmer, spelt and Polish wheat. *J. Am. Soc. Agron.* 10, 215–217.

Lisitsina, G.N. (1978) Main types of ancient farming on the Caucasus – on the basis of palaeo-ethnobotanical research. *Ber. Dtsch. Bot. Ges.* 91, 47–57.

Lucas, A. (1962) Ancient Egyptian materials and industries. 4th ed. Revised & enlarged by J.R. Harris. Edward Arnold, London.

McFadden, E.S. and Sears, E.R. (1946) The origin of *Triticum spelta* and its free-threshing hexaploid relatives. *J. Hered.* 37, 81–90, 107–116

Milisauskas, S. (1978) European prehistory. Academic Press, New York. pp. 333.

Murray, J. (1970) The first European agriculture. Univ. Press, Edinburgh.

Noy, T., Legge, A.J. and Higgs, E.S. (1973) Recent excavations at Nahal Oren, Israel. *Proc. Prehist. Soc.* 39, 75–99.

Orlov, A.A. (1923) The geographical center of origin and the area of cultivation of durum wheat *Tr. durum* Desf. Bureau of Applied Botany and Plant Breeding (Leningrad). pp. 371–459.

Pavis D'Escurac, H. (1976) La prefecture de l'Annone: Service administratif imperial d'Auguste a Constatin. Bibliotéque des Écoles francaises d'Athenes et de Rome fasc. 226 École francaise de Rome. p. 473.

Percival, J. (1921) The wheat plant. Duckworth, London. pp. 463.

Percival, J. (1926) Some new varieties of wheat. *J. Bot.* 64, 203–210.

Percival, J. (1927a) The coleoptile bundles of Indo-Abyssinian Emmer wheat (*Triticum discoccum* Schbul.) *Ann. Bot.* 41, 101–105.

Percival, J. (1927b) Wheat in 3500 BC. *Nature* 119, 280.

Pliny (1950) Natural History. Transl. by H. Rackham. Harvard University Press, Cambridge, Mass.

Riley, R. (1969) Evidence from phylogenetic relationships of the types of bread wheat first cultivated. In 'The domestication and exploitation of plants and animals'. (Eds. P.J. Ucko and G.W. Dimbleby.) Aldine, Chicago. pp. 173–176.

Schultze-Motel, J. and Kruse, J. (1965) Speltz (*Triticum spelta* L.), andere Kulturpflanzen und Unkraüter in der frühen Eisenzeit Mitteldeutschlands.. *Die Kulturpflanze* 13, 586–619.

Smith, W., Wayte, W. and Marindin, C.E. (1891) Dictionary of Greek and Roman antiquities. 2 vols. John Murray, London.

Stevens, E. (1941) Agriculture and rural life in the later Roman empire. In 'The Cambridge Economic History of Europe. Vol. I. The Agrarian life of the Middle Ages'. (Eds. J.H. Clapham and E. Power.) Cambridge University Press, Cambridge. pp. 89–117.

Stevenson, G.H. (1934) The imperial administration. In 'The Cambridge Ancient History. Vol X, (Eds. F.E. Adcock and M.P. Charlsworth.) The University Press, Cambridge. pp. 182–217.

Stewart, R.B. and Robertson III, W. (1971) Moisture and seed carbonization. *Econ. Bot.* 25, 381.

Tellez, R. and Ciferri, F. (1954) Trigos arqueológicos de Espana. Madrid Institute Nacional de Investigacion Agronómicas. p. 129.

Tumanjan, M.G. (1935) Die wildwachsenden Verwandten der kultivierten Weizen in Armenien. *Z. Pflanzenzücht.* 20, 352–363.

Van Zeist, W. (1972) Palaeobotanical Results of the 1970 Season at Çayönü, Turkey. *Helinium* 12, 1–19.

Van Zeist, W. (1976) On macroscopic traces of food plants in south-western Asia. *Philos. Trans. R. Soc. Lond. B Biol. Sci.* 275, 27–41.

Van Zeist, W. and Bakker-Heers, J.A.H. (1975). Prehistoric and early historic plant husbandry in the Altinova plain, south-eastern Turkey. In 'Final report on the excavations of the universities of Chicago, California (Los Angeles) and Amsterdam in the Keban Reservoir, Eastern Anatolia, 1968–1970. Vol. 1'. (Ed. M.N. van Loon.) Amsterdam–Oxford. pp. 223–257.

Van Zeist, W. and Bottema, S. (1966) Paleobotanical investigations at Ramad. *Ann. Archeol. Arabes Syriennes* 16, 179–180.

Van Zeist, W. and Casparie, W.A. (1968) Wild einkorn wheat and barley from Tell Mureybit in northern Syria. *Acta Bot. Neerl.* 17, 44–53.

Vavilov, N.I. (1926) Studies on the origin of cultivated plants. Institute of Applied Botany, Genetics and Plant Breeding. (Leningrad). p. 248.

Vishnu-Mittre (1974) Paleobotanical evidence in India. In 'Evolutionary studies in world crops; diversity and change in the Indian subcontinent'. (Ed. J.B. Hutchinson.) Cambridge University Press, Cambridge. pp. 3–30.

Wendorf, F., Schild, R., El Hadidi, N., Close, A., Kobusiewicz, M., Wieckowska, H., Issawi, B. and Haas, H. (1979) Use of barley in the Egyptian late paleolithic. *Science* 205, 1341–1347.

Zohary, D. (1973) The origin of cultivated cereals and pulses in the Near East. *Chromosomes Today* **4**, 307–320.

Zohary, D., Harlan, J.R. and Vardi, A. (1969) The wild diploid progenitors of wheat and their breeding value. *Euphytica* **18**, 58–65.

2   WHEAT GENETIC RESOURCES

D.R. Marshall and A.H.D. Brown

Introduction

Substantial progress in conserving the genetic resources of crop plants has been made over the last decade (Harlan, 1975; Frankel, 1978). Perhaps the most important step forward was the creation and funding of the International Board of Plant Genetic Resources (IBPGR, 1975) to sponsor and co-ordinate genetic conservation programs. In this paper, we first outline the current international programs for the genetic resources of wheat, and then discuss the more controversial scientific problems arising from these programs.

Wheat — the current situation
Organization

To facilitate the integration of the activities of the many international, regional and national institutions which are, or will become, involved in the global network of genetic resources centres, IBPGR has established Advisory Committees for each major crop. The Wheat Advisory Committee was established in 1975 and first met in 1976. CIMMYT (Centro Internacional de Mejoramiento de Maiz y Trigo) assumed co-sponsorship of the committee in 1979 and ICARDA, (International Centre for Agricultural Research in Dry Areas) which has strong research interests in durum wheats, is now also represented on the committee (IBPGR, 1980).

In contrast to the situation with some of the other major crops such as rice, there was no one single institution which could serve as a centre for wheat germplasm conservation. Consequently, the Wheat Advisory Committee recommended that four institutions be invited to establish 'base collections'. These institutions are the N.I. Vavilov Institute of Plant Industry, U.S.S.R., the USDA Agricultural Research Centre, Beltsville and Fort Collins, U.S.A., the Germplasm Laboratory of the C.N.R., Bari, Italy and the Kyoto University Germ Plasm Centre, Japan. All four institutions have now formally agreed to participate in the network. The first three are jointly responsible for the cultivated species, and the last is responsible for the maintenance of wild species. Each of the institutions has agreed to follow internationally agreed technical standards for seed storage and to guarantee the availability of material and information (Hondelmann, 1979; IPBGR, 1980).

Table 2.1
**Numbers of samples of *Triticum* sp. collected on recent expeditions
sponsored in whole or part by IBPGR (IBPGR, 1979b)**

Country or region	Number of samples
Afghanistan	200
Cyprus	105
India (N.W.)	130
Iran	111
Iraq	80
Algeria	299
Cyprus	79
Greece	95
Portugal	57
Tunisia	129
Pakistan	304
Spain	221
Syria	107

At its first meeting in 1976 the Wheat Advisory Committee, in the light of discussion at the International Symposium on Wheat Genetic Resources held at the N.I. Vavilov Institute of Plant Industry in 1975 (Brezhnev, 1976), established priorities for the exploration of wheat genetic resources. Some 13 collecting missions were sponsored in whole, or part, by IBPGR in these priority areas up until 1978 (Table 2.1).

Additional missions have been conducted in Nepal, Spain, Iraq, northern Syria and Eastern Turkey and further collecting is planned this year in Egypt, Yemen Arab Republic and Jebel Marra in the Sudan. The Committee updated its priorities in 1978, using the increased information available from individual scientists and institutions on the status of material in the field in various countries (IBPGR, 1979a). It listed 25 countries or regions where there was an urgent need for further collection; priority areas to be covered if possible over the next two to three years are summarised in Table 2.2.

The Wheat Advisory Committee, in association with an IBPGR working group on wheat descriptors, developed a list of minimum descriptors for wheat genetic resources (Table 2.3). The committee also planned to initiate a pilot evaluation program for wheat germplasm in 1979. The aim is to evaluate 200 accessions (both winter and spring) from each of three centres at 7 sites to establish the possibilities and problems of comprehensive data handling and storage.

Table 2.2
Priority areas for collection of traditional land race varieties of wheat
(IBPGR, 1979a)

Priority 1	Priority 2
Turkey – south-east and east	Spain
Caucasus – especially Armenia and Georgia	Portugal
	Yugoslavia
Syria – particularly northern areas	
	Morocco
Iraq – Zagros Mountains	
	Libya – coastal highlands and wadis
Iran – mountainous regions and eastern oases	
	Sudan – Jebel Marra region
Afghanistan – the north and mountains of Hindukush	Pakistan – several areas
Albania	Nepal – hill regions
Greece – northern Pelopponese	Bhutan
Egypt – Upper Egypt	India – several areas
Northern Yemen	P.R. China – other areas
Saudi Arabia – border areas	Mongolia
P.R. China – mountains on the west and Tibetan regions	Brazil – varieties selected for acid soils

Table 2.3
Suggested minimum descriptors for Wheat Genetic Resources (IBPGR, 1978)

Collection Data
1. Collecting Organization
2. Accession Number
3. Crop Name (wheat)
4. Genus
5. Species
6. Common/Local Names
7. Country of Origin
8. Geographic Area
9. Latitude
10. Longitude
11. Altitude
12. Donor Name
13. Donor Number
14. Location in Primary Storage – first deposit
15. Year of Primary Storage

Agronomic Evaluation
16. Days to Flower
17. Growth Habit   Winter/Spring
18. Plant Height
19. Kernel Texture   [1 = very soft;   9 = very hard]
20. Kernel Plumpness   [1 = shrivelled;   9 = plump]
21. Winter Hardiness   [1 = susceptible;   9 = resistant]
22. Drought Resistance
23. Cold Resistance
24. Sprouting Tendency   [1 = very low;   9 = high]

Morphological Evaluation
25. Stem Thickness   [1 = very thin;   9 = thick]
26. Spike Density   [1 = very lax;   9 = dense]
27. Number of Spikelets per Spike
28. Number of Kernels per Spikelet
29. Awnedness   [0/+]
30. Kernel Colour

Biochemical Evaluation Data
31. Total protein content
32. Weight of grain per unit area
33. Lysine/Protein Ratio

Collections

The number of entries held in the four proposed base collections and several regional or national base or working collections for which we could obtain published information is given in Table 2.4. Overall it is estimated that there are about 250 000 accessions of *Triticum* and related species in collections around the world with about 30% duplication of entries among the major collections (Moseman *et al.*, 1979; Hondelmann, 1979).

Table 2.4
List of major wheat collections

	Number of entries
World Base Collections	
N.I. Vavilov Institute, U.S.S.R.	21 000 (wheats)
N.S.S.L., U.S.A.	37 000 (*Triticum*)
CNR Germplasm Laboratory, Italy	26 500 (*Triticum*)
Plant Germplasm Centre, Japan	4 100 (*Triticum*)
	2 300 (*Aegilops*)
National or Regional Collections	
Gatersleben, D.D.R.	20 000 (cereals)
Braunschweig, F.R.G.	11 000 (wheats)
Australian Wheat Collection	20 000 (wheats)
ARARI, Izmir, Turkey	1 100 (*Triticum*)

As emphasised by Frankel (1977a), crop genetic resources can be grouped into four categories:
1. High yielding varieties in current use (advanced cultivars) and those they have superseded (obsolete cultivars).
2. Primitive varieties or land races of traditional, pre-scientific agriculture.
3. Wild and weedy relatives of the crop species. In the case of wheat some 25 species are included in this group (Feldman, 1979).
4. Specialised genetic and cytogenetic stocks including induced mutations.
   Unfortunately, there is little precise information on the relative representation of each of these four categories in existing collections. As a consequence, the Wheat Advisory Committee has requested the IBPGR Secretariat in FAO to undertake a detailed analysis of the different taxa and types in existing collections and of the regions where the samples were collected. This analysis, should provide quantitative information on the extent of duplication in the base collections and provide a more reliable base for the determination of collection

priorities (Hondelmann, 1979). To date the analysis has confirmed the relatively poor holdings of land-races and wild species in existing collections. For example, the maximum holding in any one collection of the *Sphaerococcum* wheats is 10 accessions and their origin is largely unknown (IBPGR, 1980).

Contentious issues in wheat germplasm conservation

As we pointed out previously (Marshall and Brown, 1975) there are definite limits to the number of accessions which can be handled effectively in programs for the conservation and utilization of genetic resources. These limits arise from the restricted resources available to carry out each stage in the overall process, illustrated by the following scheme:

It is important, therefore that these limited resources should be used in the best possible way. While there is general agreement on the need for efficient procedures for the collection, classification and description (particularly with respect to the detection of duplicate accessions), preservation, evaluation and utilization of genetic resources, there is far less agreement as to how these needs should be met. The problems associated with the development of optimum operating procedures are exacerbated in wheat as four centres are involved in the maintenance of base collections.

Exploration – optimum sampling procedures

One of the more controversial issues in genetic conservation is the definition of optimal sampling strategies. The use of efficient sampling procedures is clearly important in the case of traditional land race varieties, particularly where such populations are threatened by extinction, or occur in remote areas or difficult terrain, so that they are likely to be sampled only once. In these cases, the limits on the number of samples which can be collected are severe (Bennett, 1970) and plant explorers carry a great responsibility because it is their decisions which determine to what extent the gene resources in such populations will be available for the use of future generations. As emphasized by Allard (1970) the problem facing the plant explorer is how to collect the maximum amount of useful genetic variability and still keep the number of samples within practical limits.

In order to develop a quantitative theory of sampling it is necessary to define (i) an appropriate measure of genetic diversity, the parameter that is to be maximised, and (ii) 'useful' genetic variation which is to have priority in sampling. We argued previously (Marshall and Brown, 1975) that the number of alleles per locus provides the best measure of genetic diversity in the context of genetic conservation. First, it is a direct measure of genic variation in a population. It is more reliable, therefore, than diversity indices based on variance in quantitative characters which measure only that portion of genetic variability expressed phenotypically. Second, it is the measure of most concern to the plant explorer. Genetic conservationists are interested in preserving at least one copy of each variant in a population irrespective of their relative frequencies. The number of allelic variants of interest in a sample compared to the number in the population is, therefore, the critical index of sampling efficiency for genetic conservation. It should be emphasized that, in the present context, combinations of alleles at two or more loci can be regarded as 'alleles' at a 'super gene'. Thus, while we frame our theory in terms of alleles at a single locus, the same theory applies to genotypes at several loci including coadapted gene complexes and morphological variants for qualitative or quantitative characters governed by two or more genes.

Since the aim of genetic conservation is to ensure the availability of adequate germplasm for immediate and future use by plant breeders, and further, since future germplasm needs are unpredictable, no one class of loci or traits deserves priority in sampling. Rather, the aim should be to preserve as much variation as possible at all loci. Thus, in defining 'useful' genetic variation which deserves priority in sampling, we need consider only the different classes of alleles which occur at each locus. Allelic variants can be divided into four classes depending on their population frequency and distribution in the target plant (Marshall and Brown, 1975, 1980). First, the variants in each population can be classed into those which are *common* (frequency greater than, say, 5%) and those which are *rare* (frequency less than 5%). Next each variant with at least one common occurrence is classed as to whether it is common *locally* (that is, common in one or a few adjacent populations) or *widespread* (common in many populations). Similarly variants which are rare can be classed into whether they are *localised* or *widespread*. Of these four classes, it is the *locally common* alleles which take priority in sampling. Such alleles are more difficult to collect than either *widespread common* alleles, which will inevitably be included in the sample regardless of strategy, or *widespread rare* alleles, whose numbers in a sample will depend largely on the total number of individuals collected rather than how those individuals are distributed between and within sample sites. Further, common alleles presumably represent adaptive variants maintained in populations by some form of balancing selection (Dobzhansky, 1970) and are therefore of great interest to plant breeders. On the other hand, *locally rare* alleles are of less interest to plant breeders and are far more difficult to collect than their *locally common* counterparts, and hence do not merit priority in sampling.

In the light of the above the aim of plant exploration for genetic conservation is the collection of at least one copy of each variant occurring in any population of the target species with frequency greater than 0.05. Assuming the collector has no *a priori* knowledge of the distribution of variation in the target populations the optimum strategy to achieve this objective is:

(i)   to collect 50 to 100 individuals per site at random,

(ii) to sample as many sites as possible within the time available and

(iii) to ensure that sampling sites represent as broad a range of environments as possible. If some information is available on the distribution of variation in nature, as would be the case if the two stage sampling procedure recommended by Jain (1975) were followed, then more sophisticated sampling strategies can be adopted (see Marshall and Brown, 1975 for details). We have also outlined various modifications to the above strategy which may be necessary when difficulties in the field prevent the use of the optimum procedure (Marshall and Brown, 1980).

The above optimum strategy has been criticised on a number of grounds. Qualset (1975), for example, argued that conserved genetic resources serve both the plant breeder and the geneticist studying the evolution of characters and character complexes in cultivated species. To ensure that interesting, but rare, combinations of characters are obtained in samples taken from agricultural populations, he argued that the sample size should be much larger, about 500 individuals per population rather than 50. He acknowledged that because of the greater resources needed to obtain and maintain these larger samples, his recommendation was not practical in general application. He suggested, therefore, sample sizes of 50–100 individuals be taken from most populations, but that a proportion of the populations, say 10% should be sampled more extensively.

The provision of material for evolutionary and ecological studies of crop plants is an important function of collections and large samples are needed to study character or gene combinations. In fact, even a sample of 500 individuals per population may not be adequate. For example, consider the case of a specific allelic combination at two loci which occurs in a population with frequency .0025. To be 95% sure of obtaining one such allelic combination requires a sample size of 1200 individuals. To be 95% sure that such an allelic combination is represented in the sample with frequency .0025 ± .001% requires a sample size of about 10 000 individuals. The point is that sampling for evolutionary studies, where one is interested in obtaining an accurate representation of the allelic frequencies in the population, requires the collection of much larger numbers of individuals, than sampling for genetic conservation (Marshall and Brown, 1980). Further, compromise strategies such as that suggested by Qualset tend to reduce the efficiency of sampling for genetic conservation without meeting the needs of the evolutionists. We would argue that samples of 50–100 individuals should be collected from as many sites as possible in a region and that evolutionary studies should be undertaken on the pooled sample. If the rare variants of interest are widespread then they will be equally well represented in a combined sample as in a single large sample from one population. If the rare variants occur in only one or few populations, the probability that each will be collected is very small unless very large samples are taken from all populations.

Witcombe and Gilani (1979) criticised our approach on the basis of their studies on variation in landrace populations of wheat and barley from the Himalayas. They claimed that our theory of collection "is based on isozyme variation i.e. individual alleles", whereas "plant genetic resources are not exploited for isozyme variation but for agronomically useful characters". We certainly argued that the number of alleles was the best direct measure of genetic diversity in terms of framing sampling strategies for the reasons outlined above. Further, we used the data available from isozyme studies, as well as theoretical studies, to illustrate the range of allelic distributions at individual loci which may be encountered in practice. It was then shown that a sample of 50–100 individuals was adequate to ensure the collection of all common alleles in a population for virtually all such distributions. However, this does not mean that our sampling technique applies only to isozyme variants. Indeed, as we stressed previously, it applies to all loci. Genetic variation in quantitative traits, including agronomically important ones, is due to allelic variation at loci governing these traits. However, the effects of the individual loci or alleles are not morphologically distinguishable. Nevertheless, the aim of plant exploration is still to collect as many alleles at each locus as possible, as this maximizes the plant breeder's capacity to manipulate these characters in breeding programs. Second, while there has been little effort to date to select directly for enzyme variants, selection for agronomically important characters may often result in indirect selection for biochemical variants, some of which will be electrophoretically detectable. As our knowledge of physiological and biochemical processes underlying agronomically important characters increases, it will be possible to select directly for such variants (Hageman, *et al.*, 1967).

More recently, Bogyo *et al.*, (1980) were also critical of some aspects of our sampling strategy on the basis of their analysis of the patterns of variation in Sicilian landraces of durum wheat. On the one hand, they questioned the validity of random sampling from phenotypically heterogeneous populations. They estimated that in their collections 94% of all recognizable spike types (68 out of 72) were 'rare' because they occurred with frequency less than 0.05% in the entire collection. They argued that rare and unusual morphological variants are intrinsically more interesting to the explorer than common variants and that much larger sample sizes than we recommended are required to collect such variants, unless they were preferentially sampled by the collector. On the other hand, from estimates of the components of variation for quantitative characters within and among samples and regions, they calculated the optimum sampling procedure for such traits using Cochran's (1963) multistage sampling theory. These calculations suggested that our recommended sample size of 50–100 individuals was far too large. They concluded that "it is almost impossible to make a general recommendation for (an) optimum sampling strategy" and further, "sampling strategy should be different for characters which are inherited quantitatively from those which are controlled by one or two genes". We disagree with these conclusions.

The use of biased sampling of rare phenotypic variants had been advocated earlier by Bennett (1970) and we considered its validity in some detail in our 1975 paper. While we agree with Bogyo *et al.* (1980) that it is difficult for collectors not to bias their samples in favour of types which are new and unusual to them, we believe that the conscious and deliberate selection of rare types has little to recommend it as a general policy. Biased sampling is more time consuming than random sampling because of the need to search through large numbers of plants for the rare types. By collecting rare types at one site, the explorer invariably reduces his opportunities of collecting locally common variants at other sites and as we pointed out above, common, currently adaptive alleles are more likely to be valuable to the plant breeder than rare variants. Further, if only biased samples are collected, this will severely reduce their value for evolutionary studies. If both a random sample and a biased sample are collected, this will increase the effort which must be devoted to all other aspects of conservation, evaluation and utilization.

It is worth noting that Bogyo *et al.* (1980) used a different definition of a common *vs* rare allele. We defined a rare allele to be one with a frequency less than 0.05 *within a population*. They defined a rare allele to be one with a frequency less than 0.05 in their *total collection* of some 12 000 spikes. In consequence 94% of spike types were rare by their definition and this led them to place undue emphasis on the collection of rare types. Yet, it is possible that many of these spike types were *locally common*, and occurred in one or a few populations sampled with frequency greater than 0.05. Alternatively they may have occurred at low frequency in a great many populations. The fact that they collected 68 spike types which were rare in the region as a whole using the procedure we recommended illustrates its efficiency in recovering locally common or widespread rare variants.

The conclusions of Bogyo *et al.* that much smaller samples are required adequately to sample variaton at loci governing quantitative traits could arise from the fact that variance in such characters does not measure the number and distribution of alleles at the loci responsible for such characters. In general, there is no reason why alleles at quantitative loci should have distinctly different distributions to alleles at qualitative loci, and hence, why the optimum sample sizes to collect all common alleles should be different for the two classes of characters.

Overall, despite the criticisms of the sampling strategy developed in our 1975 paper, we believe our conclusions are still valid.

Classification and description − the identification and elimination of duplicate samples

One issue which has received only limited attention in genetic conservation programs is the identification and elimination of duplicate entries within and among collections. However, this issue is an important one and it will demand greater attention in the future because of the increased level of exploration in recent years and the need to rationalize and consolidate existing collections. This issue is of particular importance in wheat because of the large number of existing collections and their high degree of redundancy.

Recognizing and reducing redundancy within a collection

The first step in eliminating duplicate entries is the recognition of the duplicates. Hondelmann (1979) has suggested that the published list of minimum descriptors (Table 2.3) be used for this purpose. However, this list of descriptors was apparently developed for two quite different purposes − identification and evaluation. As a consequence, the list contains many descriptors, particularly those difficult to evaluate objectively (e.g. drought and cold tolerance, winter hardiness, and sprouting susceptibility), or expensive to measure, (e.g. protein content and percent lysine/unit protein) which are relevant to evaluation but which have little to recommend them in terms of the detection of duplicates. Further, the list omits many simply recorded morphological characteristics such as awn colour, persistence and barbing, glume colour and hairiness, and biochemical characteristics such as electrophoretic patterns, which could be readily used to identify duplicate entries (Jain, 1979).

We suggest that separate specialized descriptor lists should be developed for the identification and characterization of entries on one hand, and evaluation on the other. Further, the recording of the first set of descriptors should be the responsibility of the collection curators while plant breeders, and not the curators, should be responsible for evaluation. We discuss the point in more detail below.

Given a sufficient number of characters are scored, then detection of duplicates among fixed homozygous lines and varieties would be a relatively simple matter. However, identifying duplicates among population samples from highly heterogeneous land race varieties is a more difficult problem. Frankel and Soulé (in press) have suggested that multivariate analyses of all available data, including location data (site, region, country) and measurement data for quantitative characters and enumeration data for qualitative characters, should provide presumptive evidence of redundancy and diversity amongst samples. However, the use of multivariate analysis requires that some weighting be given to each trait. This raises a number of questions:

(a)   If two entries are known to come from different sites, is this presumptive evidence in itself that they are different? If not, is the site, the region or the country of origin any value at all in determining redundancy?

(b)   With respect to quantitative traits, should measurements be made on bulked samples to estimate only mean values or on an individual plant basis to estimate both means and variances? Are two entries to be regarded as sufficiently distinct to warrant retention as separate entries if both means and variances are significantly different, or if the means differ and the variances are similar, or if variances differ and the means are similar? In other words — what relative weights should be given to means and variances for quantitative traits in determining redundancy?

(c)   With respect to qualitative traits, should a sample be regarded as redundant only if the same set of variants occur in another popular at similar frequencies, or if it has the same set of variants as another population regardless of frequency?

(d)   What relative weights should be given to qualitative and quantitative characters in determining redundancy? Are quantitative characters better indicators of population differences than qualitative characters, or vice-versa?

With reference to the questions in (a) above, landrace samples from different ecological areas or regions, such as those described by Bogyo et al. (1980), are usually assumed to be distinct and worthy of preservation. If this is done the problem reduces to one of determining if one, several or all samples collected from within a region are distinct. Further, in response to (b) above, we would suggest that populations which differ in either the mean or variance or both should be regarded as distinct. The reason is that both the mean and variance are indirect and often unreliable indicators of genic diversity at loci controlling quantitative traits. Thus, two populations with different means or variances may in fact contain the same alleles at underlying loci at different frequencies. On the other hand, populations with the same mean or variance may carry different alleles at these loci. In consequence, when dealing with quantitative traits it would seem reasonable to follow a conservative strategy and assume redundancy only if both the means and variances of samples are similar.

With respect to (c) above we would argue that two populations could be regarded as equivalent if they contain the same or closely similar sets of variants or alleles for qualitative characters. The reason is that the breeder can presumably recover the variants if necessary from the population regardless of frequency. Further, two samples drawn from the identical base populations will often differ in gene frequencies at marker loci due simply to sampling error.

Finally in response to (d), we suggest that qualitative traits provide better discrimination between samples than quantitative traits simply because the variants at loci governing qualitative characters are distinguishable and provide a more direct measure of genic diversity. Nevertheless, there are dangers in relying completely on one particular class of character in estimating genetic diversity within and among samples. As Kahler *et al.* (1980) concluded from an extensive series of studies on the patterns of genetic variation in natural populations of slender wild oat (*A. barbata*) "it is apparent that no single class of loci, such as those governing enzyme variants, or those governing morphological polymorphisms, or those governing measurement characters, gives a complete picture of the extent of genetic variability within populations or of degree of evolutionary divergence among populations in this species".

Once it has been decided which entries are redundant, the next problem is to decide what steps should be taken to reduce redundancy. This problem has been recently considered in detail by Frankel and Brown (in Frankel and Soulé, in press). They suggested four possible courses of action:

(a) Retain all entries regardless of apparent redundancy. It was suggested that this would be appropriate for collections of primitive cultivars where there was little likelihood of further exploration, or for collections of great scientific and historical interest.

(b) Retain a random selection of redundant samples.

(c) Bulk the redundant entries and retain a subsample. This would result in a single bulk accession with consequent savings in storage, evaluation and regeneration costs.

(d)Reduce the sample size of all entries but continue to maintain them separately.

Options b, c and d lead to a loss of genetic variation compared to (a). Brown (in Frankel and Soulé, in press) examined the extent of this loss in each case using simple probability models. On the basis of these calculations he concluded that (b) above, the random deletion of entries, would seldom if ever be an appropriate course of action. However, there was little difference between (c) and (d) in terms of loss of genetic variation for a given reduction in total sample size. Brown argued therefore that the choice between (c) and (d) must be made on grounds (relative cost, difficulty of detecting rare variants in large bulks, etc.) other than efficiency in retaining genetic variation.

Reducing redundancy among collections

The basic procedures for the recognition and reduction of redundancy among collections are the same as within a collection. However the problem of recognizing redundant entries is likely to be exacerbated if they are held by different institutes. Either an intensive effort will be needed to ensure standardisation of data collection among institutes or one or several institutes will need to collect data on the total collection.

Further there is likely to be much less impetus to reduce redundancy among collections. If a national or regional gene bank has a redundant entry, it may be reluctant to discard it even though it may be identical to material held in another gene bank. This is particularly true if the entry may have to be reintroduced through quarantine at some future date. Indeed, there may be considerable pressure, given quarantine difficulties and present political uncertainties, for redundancy to increase, rather than decrease, among collections. For example, if a breeder introduces and evaluates a series of lines not held by his regional or national collection from overseas, then there would be strong arguments to have this material added to his own regional collection for future reference. If this process is repeated continuously then there would quickly be 100% redundancy among established collections.

Conservation – the role of dynamic conservation in wheat

Most, if not all, the current programs for the preservation of the genetic resources of the cultivated wheats and their wild relatives are based on 'static' conservation (Frankel and Soulé, in press); that is, on the preservation of seeds in low temperature, low humidity, stores. However, this policy of sole reliance on 'static' preservation has been questioned repeatedly in the literature over the last 15 years. As an alternative it has been suggested that emphasis should be given to 'dynamic' conservation methods i.e. those allowing continuing evolution

(Frankel and Soulé, in press). Three such 'dynamic' conservation methods have been proposed: (i) preservation of land races in their areas of cultivation (i.e. conservation of land races *in situ*, Bennett, 1968); (ii) the use of mass reservoirs or composite crosses with a large number of diverse parents (Simmonds, 1962) and; (iii) genetic reserves (Dinoor, 1975) or gene parks (Browning *et al.*, 1979). It has been alleged that such dynamic conservation methods would permit the efficient preservation of existing variants and, in addition, allow scope for the emergence of new variants. Advocates of dynamic conservation methods argue that the need for continuing evolution is greatest in the case of genetic variability for disease and pest resistance. Their argument proceeds as follows. The present genetic variability for disease resistance in land race populations of wheat has developed as a result of the continued co-evolution of the host and its parasites. 'Static' conservation of such populations stops adaptive evolution in the host. But evolution in the parasites can continue unhindered on the cultivated species and its wild relatives. The net result is that the breeder will ultimately end up with limited finite resources for resistance in the host to counter potentially unlimited and evolutionarily labile resources for virulence in parasites.

Various authors have discussed the role of 'dynamic' conservation methods in preserving existing variation in cultivated crop species (Frankel, 1970a, 1978; Frankel and Soulé, in press; Marshall and Brown, 1975). From these discussions it is evident that as far as preserving existing variation is concerned, 'dynamic' conservation has few, if any, advantages over 'static' conservation. However, the question of the need for 'dynamic' conservation as a supplement to 'static' conservation particularly in the case of disease resistance has received less attention. This is despite the fact that the issue is a valid one of considerable practical importance, and the three proposed methods of dynamic conservation potentially meet this need.

The question we have to answer in this regard is — will the cultivation of heterogeneous populations of cultivated wheats (land race populations *per se*, mixtures of land races or bulk hybrid populations) grown in centres of diversity for specified pathogens, provide adequate potential for the evolution of new resistance sources? It is of course difficult to provide a definitive answer to this question because of a lack of knowledge on how new sources of resistance evolve. However, the potential of these methods in providing future plant breeders with effective genes for resistance seems limited simply because of the limited number of parents which could be included and the limited size of the populations which could be grown. It is unlikely that more than a few hundred hectares could be devoted on a world scale to the cultivation of heterogeneous populations for continuing evolution. The evolutionary potential of such

populations would be relatively modest compared to the many land race populations, with their enormous diversity, grown over millions of hectares, which
formerly provided for the evolution of resistance.

The only alternative appears to be to ensure that the wild relatives of the
cultivated wheat are adequately conserved *in situ*, in nature reserves within a
co-adapted community, where this is possible. Dynamic conservation of wild
species in such reserves would satisfy the requirement for continuing evolution
without the continuing need to cultivate extremely large populations. Of
course, the utilization of newly evolved resistances would be more difficult in
case of wild species. However, this disadvantage would be more than offset by
the advantages that populations of wild species are self regenerating and would
presumably occupy relatively large areas.

Evaluation and utilization — does evaluation limit utilization?

Evaluation remains a continuing problem area in genetic conservation, particularly in wheat. As has been emphasized by Frankel (1970b) evaluation is an
essential preliminary to utilization, and further, the more information that is
available, the more valuable the collection. In addition, Frankel (1977b) has
argued "It is no overstatement that the relatively limited use which has been
made of large existing collections is mainly due to the deficiencies of existing
documentation, which may restrict the usefulness of a collection even within the
institution owning it and much more so to other potential users. The difficulty
arises from the vast amount of information and from the lack of agreed standards and procedures which would facilitate the communication of information
between collections and between collections and users".

This philosophy has gained wide credence in recent years. As a consequence,
considerable effort and substantial resources have gone into (i) the definition of
agreed standards and procedures for evaluation. In the case of wheat, this effort
has led to the publication of the recommended list of minimum descriptors
(IBPGR, 1978); (ii) the development of international schemes for evaluation of
world germplasm collections. Again, for wheat, a pilot scheme was recommended in 1976 and initiated in 1979 for the evaluation of 200 lines from each of
three collections at seven sites. Presumably if this scheme is successful then it
will be extended to cover all accessions. If each of the approximately 200 000
accessions in the four base collections is evaluated for the 33 descriptors at each
of 7 sites this will generate a data bank of about 46 million records; (iii) the
development of information storage and retrieval systems to handle the large
body of data to be collected (e.g. IBPGR, 1977).

One may ask whether the resources which have been devoted to the development of such centralized evaluation/documentation systems, and the resources

which will need to be devoted to their completion, are justified. Will the availability of these large data banks enhance the utilization of genetic resources in general and those of wheat in particular? The answers to these questions, we would suggest, are negative.

Several of the premises on which the development of centralized co-operative evaluation programs are based are questionable. First, while it is true that evaluation is a necessary first step to utilization, it does not necessarily follow that lack of information has been a major stumbling block to the greater utilization of genetic resources in the past. Indeed, it can be argued that it is the common belief of plant breeders, based on experience, teaching or hearsay, that collections represent good sources of specific characteristics, such as disease resistance, and little else (Simmonds 1962), which has led to a narrowing of the effective parental gene pool in many breeding programs (Frankel, 1954; Frankel, 1970b; Day, 1973). Swamping breeders with a mass of information much of which is irrelevant to their breeding efforts is hardly likely to change this belief. For instance how many wheat breeders in the world are breeding for improved lysine/unit protein? What is needed are more studies of the likely benefits which may derive from the use of a broader germplasm base in plant breeding, and how these benefits can best be achieved. Studies such as those of Frey and his colleagues (Lawrence and Frey, 1975; Takeda and Frey, 1976, 1977), which indicated that transgressive segregates for high grain yield occurred frequently in crosses between wild and cultivated oats resulting in yield increases of 27–30%, are more likely to promote interest in the utilization of diverse germplasm than mere routine standardized evaluation.

Second, it is assumed that co-ordination by conservation centres will increase efficiency of evaluation. Yet the breeders will still have to grow the material in their own fields to assess 'location specific' traits not included in the minimum descriptor list. Further, most breeders like to assess material in their own environments for all characters. It could be argued on this basis that current proposals will enhance rather than diminish duplication of effort in evaluation. We would suggest that conservation centres should be concerned solely with description of material in terms of its origin and descriptive characteristics, particularly those of use in detecting redundancy, and evaluation should be the responsibility of the users. The diverse aims of breeding programs and the location-specific nature of many characteristics of importance in breeding, raises doubts as to the relevance of any list of characters which claims to be of general or universal interest to breeders.

Third, the premise that the development of large computerized information networks will increase the flow of information among breeders is open to serious doubt. Breeders have always had the opportunity to publish the results of their evaluation studies through normal channels — scientific papers and reports. Some have chosen to use these channels (e.g. Vogel *et al.*, 1973, 1976; Porceddu, 1976), some have not. We would ask if those who do not take the time to write up their results for publication are any more likely to provide a summary of these results to a centralized information network. Further, in seeking information on variation in a collection for some trait of interest, is a breeder more likely to consult a scientific paper where the results have been summarized and analyzed or an undigested list of data points on a computer print-out?

In short, we would argue that while evaluation and utilization are directly linked, utilization is the ultimate goal of genetic conservation. Further, evaluation of more lines for more characters at more sites will not automatically promote greater utilization of collections. With the current emphasis on centralized co-ordinated evaluation/documentation systems there is a danger that evaluation, and the manipulation of the large bodies of data generated, and not utilization will be seen as the primary goal of these systems. It is important that this danger be avoided and that utilization of collections be fostered along realistic lines or much of the motivation for their continuing existence will be lost.

References

Allard, R.W. (1970) Population structure and sampling methods. In 'Genetic Resources in Plants — their Exploration and Conservation'. (Eds. O.H. Frankel and E. Bennett.) Blackwell, Oxford and Edinburgh. pp. 97–107.

Bennett, E. (1968) Record of the FAO/IBP Technical Conference on the Exploration, Utilization and Conservation of Plant Genetic Resources, 1967. FAO, Rome.

Bennett, E. (1970) Tactics of plant exploration. In 'Genetic Resources in Plants — their Exploration and Conservation'. (Eds. O.H. Frankel and E. Bennett.) Blackwell, Oxford and Edinburgh. pp. 157–179.

Bogyo, T.P., Porceddu, E. and Perrino, P. (1980) Analysis of sampling strategies for collecting genetic material. *Econ. Bot.* **34**, 160–174.

Brezhnev, D.D. (1976) Wheat Genetic Resources — Proceedings of an International Symposium. Leningrad, U.S.S.R.

Browning, J.A., Frey, K.J., McDaniel, M.E., Simons, M.D. and Wahl, I. (1979) The bio-logic of using multilines to buffer pathogen populations and prevent disease loss. *Indian J. Genet. Plant Breed.* **39**, 3–9. Discussion pp. 105–106.

Cochran, W.G. (1963) Sampling techniques. (2nd edition). Wiley, N.Y.

Day, P.R. (1973) Genetic variability of crops. *Annu. Rev. Phytopathol.* **11**, 293–312.

Dinoor, A. (1975) Evaluation of sources of disease resistance. In 'Crop Genetic Resources for Today and Tomorrow'. (Eds. O.H. Frankel and J.G. Hawkes.) Cambridge University Press, Cambridge. pp. 201–210.

Dobzhansky, Th. (1970) 'Genetics of the Evolutionary Process.' Columbia University Press, New York.

Feldman, M. (1979) Genetic resources of wild wheats and their use in breeding. *Monografia Genetica Agraria* 4, 9–26.

Frankel, O.H. (1954) Invasion and evolution of plants in Australia and New Zealand. *Caryologia* VI (Suppl), 600–619.

Frankel, O.H. (1970a) Genetic conservation in perspective. In 'Genetic Resources in Plants – their Exploration and Conservation'. (Eds. O.H. Frankel and E. Bennett.) Blackwell, Oxford and Edinburgh. pp. 469–490.

Frankel, O.H. (1970b) Evaluation and utilization – Introductory remarks. In 'Genetic Resources in Plants – their Exploration and Conservation'. (Eds. O.H. Frankel and E. Bennett.) Blackwell, Oxford and Edinburgh. pp. 395–402.

Frankel, O.H. (1977a) Natural variation and its conservation. In 'Genetic Diversity in Plants'. (Eds. A. Muhammed, R. Aksel and R.C. von Borstel.) Plenum. pp. 21–44.

Frankel, O.H. (1977b) Genetic resources. *Ann. N.Y. Acad. Sci.* 287, 332–344.

Frankel, O.H. (1978) Conservation of crop genetic resources and their wild relatives: an overview. In 'Conservation and Agriculture'. (Ed. J.G. Hawkes.) Duckworth, London. pp. 123–149.

Frankel, O.H. and Soulé, M. (1980) 'Conservation and Evolution.' Cambridge University Press (in press).

Hageman, R.H., Leng, E.R. and Dudley, J.W. (1967) A biochemical approach to corn breeding. *Adv. Agron.* 19, 45–86.

Harlan, J.R. (1975) Our vanishing genetic resources. *Science* 188, 618–621.

Hondelmann, W. (1979) Wheat germplasm collection conservation and utilization through international cooperation. Proc. 5th Int. Wheat Genetics Symp. New Delhi, 1978. Vol. 7, pp. 149–155.

IBPGR (1975) 'The Conservation of Crop Genetic Resources.' The Whitefriars Press Ltd., London. 15 pp.

IBPGR (1977) Report of the second meeting of the Advisory Committee of IBPGR on the Genetic Resources Communication, Information and Documentation System (GR/CIDS). IBPGR Secretariat, Rome.

IBPGR (1978) Descriptors of wheat and Aegilops. IBPGR Secretariat, Rome.

IBPGR (1979a) Annual Report, 1978. IBPGR Secretariat, Rome.

IBPGR (1979b) A Review of Policies and Activities 1974–78 and the Prospects for the Future. IBPGR Secretariat, Rome.

IBPGR (1980) Annual Report, 1979. IBPGR Secretariat, Rome.

Jain, S.K. (1975) Population structure and the effects of breeding system. In 'Crop Genetic Resources for Today and Tomorrow'. (Eds. O.H. Frankel and J.G. Hawkes.) Cambridge University Press, Cambridge. pp. 15–36.

40   *D.R. Marshall and A.H.D. Brown*

Jain, S.K. (1979) Biosystematic studies of populations in germplasm collections. Proc. Conf. Broadening of the Genetic Base Crops, Wageningen. Pudoc, Wageningen.

Kahler, A.L., Allard, R.W., Krzakowa, M., Wehrhahn, C.F. and Nevo, E. (1980) Associations between isozyme phenotypes and environment in the slender wild rat (*Avena barbata*) in Israel. *Theor. Appl. Genet.* **56**, 31–47.

Lawrence, P.K. and Frey, K.J. (1975) Backcross variability for grain yield in oat species crosses (*Avena sativa* L. x *A. sterilis* L.). *Euphytica* **24**, 77–85.

Marshall, D.R. and Brown, A.H.D. (1975) Optimum sampling strategies in genetic conservation. In 'Genetic Resources for Today and Tomorrow'. (Eds. O.H. Frankel and J.G. Hawkes.) Cambridge University Press, Cambridge. pp. 53–80.

Marshall, D.R. and Brown, A.H.D. (1980) Theory of forage plant collection. In 'Genetic Resources in Forage Plants'. (Eds. R.A. Bray and J.G. McIvor.) CSIRO, Melbourne. (In press.)

Moseman, J.G., Kilpatrick, R.A. and Porter, W.M. (1979) Evaluation and documentation of pest resistance in wheat germplasm collections. Proc. 5th Int. Wheat Genetics Symp. New Delhi, 1978. Vol. 1, pp. 143–148.

Porceddu, E. (1976) Variation for agronomical traits in a world collection of *durum* wheat. *Z. Pflanzenzücht.* **77**, 314–328.

Qualset, C.O. (1975) Sampling germplasm in a centre of diversity: an example of disease resistance in Ethiopian barley. In 'Crop Genetic Resources for Today and Tomorrow'. (Eds. O.H. Frankel and J.G. Hawkes.) Cambridge University Press, Cambridge. pp. 81–96.

Simmonds, N.W. (1962) Variability in crop plants, its use and conservation. *Biol. Rev.* **37**, 442–465.

Takeda, K. and Frey, K.J. (1976) Contribution of the growth rate and harvest index to grain yield of progenies from *Avena sativa* and *A. sterilis* crosses. *Crop Sci.* **16**, 817–821.

Takeda, K. and Frey, K.J. (1977) Growth rate inheritance and associations with other traits in backcross populations of *Avena sativa* x *A. sterilis. Euphytica* **26**, 309–317.

Vogel, K.P., Johnson, V.A. and Mattern, P.J. (1973) Results of systematic analysis for protein and lysine composition of common wheats (*Triticum aestivum* L.) in the World Wheat Collection. *Nebr. Agric Exp. Stn. Res. Bull.* **258**.

Vogel, K.P., Johnson, V.A. and Mattern, P.J. (1976) Protein and lysine content of grain, endosperm and bran of wheats from the USDA World Wheat Collection. *Crop Sci.* **16**, 655–660.

Witcombe, J.R. and Gilani, M.M. (1979) Variation in cereals from the Himalayas and the optimum strategy for sampling plant germplasm. *J. Appl. Ecol.* **16**, 633–640.

3 MOLECULAR ASPECTS OF WHEAT EVOLUTION: REPEATED DNA SEQUENCES

W. J. Peacock, W. L. Gerlach and E. S. Dennis

Polyploid wheats are of two basic types — hexaploids with 21 pairs of chromosomes and tetraploids with 14 pairs of chromosomes. Cytogenetic analysis has established that the two genomes (A and B) of tetraploid wheat are different but homoeologous. Each has 7 pairs of chromosomes. These two genomes occur in the hexaploid wheats together with an additional genome (D), again of 7 pairs of chromosomes. The different genomes have become associated through hybridization events between different diploid species of grasses, the progenitors of the modern wheats.

Geneticists have been interested in identifying the diploid species which hybridized to form polyploid wheats not only because the evolution of cultivated wheats is closely associated with the social evolution of man, but also because knowledge of the progenitor species could be of importance in facilitating the introduction of genes into modern agricultural cultivars. Furthermore, the transition from diploid to tetraploid and hexaploid chromosome levels provides us with an opportunity to analyse and document changes that occur in plant genomes under the intense selection pressures associated with the development of agricultural crops.

The D genome of the hexaploid wheats is thought to have been derived from the diploid goat grass *Aegilops squarrosa*. A synthetic hexaploid formed by crossing *Ae. squarrosa* with the tetraploid wheat *Triticum dicoccoides* was morphologically indistinguishable from *T. aestivum* ssp. *spelta*, a naturally occurring hexaploid (McFadden and Sears, 1946). Meiotic behaviour in the hybrid between the synthetic and naturally occurring hexaploids was regular, indicating the equivalence of the *Ae. squarrosa* and D genomes. The origin of the other two genomes is by no means as certain. Morphological characters, isozyme variants and chromosome pairing characteristics have all been used in attempts to determine which, if any, of the diploid wheats or related grasses were the source of either the A or B genomes. The consensus is that the A genome was derived from a diploid *Triticum* species, the most favoured being the cultivated *T. monococcum* or the wild *T. boeoticum*. The identity of the B genome donor is very much an open question. Many different species have been proposed (Table 3.1) but none has all of the B genome properties, and different criteria have given conflicting indications.

Table 3.1

Species which have been proposed as donors of the B genome to tetraploid wheats and the basis for their proposition

Species	Basis of proposition	Reference
Aegilops speltoides	Spikelet morphology Chromosome pairing and karyology Esterase isozyme spectra Nuclear DNA content Nuclear-cytoplasmic interactions	Sarkar and Stebbins (1956) Riley *et al.*, (1958) Jaaska (1980) Rees and Walters (1965) Suemoto (1978)
Ae. bicornis	Amphiploid with *T. monococcum* resembles tetraploid wheat	Sears (1956)
Ae. longissima	Amylase inhibitor spectrum Nuclear DNA content	Vitozzi and Silano (1976) Nishikawa and Furuta (1978)
Ae. searsii	Chromosome pairing and karyology	Feldman (1978)
Triticum urartu	Seed protein electrophoretic profiles Anther morphology Spike and spikelet characters	Johnson (1975) Johnson and Dhaliwal (1978) Dhaliwal and Johnson (1976) Dhaliwal (1976)
Agropyron	Morphology	McFadden and Sears (1946)
Polyphyletic origin	Absence of a single demonstrable donor and the occurrence of introgression in polyploid plant series C-banding	Zohary and Feldman (1962) Natarajan and Sharma (1974)

It is possible that in seeking the identity of the wheat ancestors, both the methodologies and the questions have been inadequate. For example, meiotic chromosome pairing, which in many circumstances is a powerful indicator of chromosome homology, can be acutely affected by the action of modifying genetic elements (Riley and Chapman, 1958; Sears and Okamoto, 1958; Dover and Riley, 1972). Another potential problem is that the genomes as they now occur in the cultivated wheats may not reflect a single hybridization event. Additional hybridizations involving the initial amphiploids could lead to introgression and partial substitution of the original genomes (Zohary and Feldman, 1962).

We have approached the problem of the evolution of the modern wheats by looking at the molecular and chromosomal organisation of repeated DNA sequences isolated from the hexaploid genome. By examining segments of the chromosomal DNA molecules themselves, we avoid some of the uncertainties mentioned above and are able to trace particular chromosomal regions rather than genomes as a whole.

Repeated DNA sequences as markers of evolutionary events

DNA sequences which are repeated many times in the genome of hexaploid wheat can be detected in other polyploid wheats, related cereals, and in the genomes of wild diploid wheat species (Flavell *et al.*, 1977, 1979). The fact that these DNA sequences are identifiable over a wide range of species suggests that it should be possible to use the chromosomal distribution pattern of a sequence to trace the evolutionary history of chromosomes and chromosome segments.

We have isolated highly repeated sequences from hexaploid wheat and mapped their chromosomal locations (Gerlach and Peacock, 1980). The sequences were isolated as rapidly renaturing DNA, and used as templates for the production of radioactive complementary RNA which was hybridized *in situ* to metaphase chromosome preparations. Autoradiographs showed that these sequences are located primarily on the seven chromosomes of the B genome, with major sites also on chromosomes 4 and 7 of the A genome. Each of these nine chromosomes has a distinctive pattern of sites and can be recognised in both tetraploid and hexaploid wheats, suggesting that these chromosomes might be recognisable in diploid wheat species. The concentration of highly repeated DNA in the B genome of hexaploid wheat also suggested that any B genome donor would be readily differentiated from A or D genome ancestors. The diagnostic patterns of sites on two chromosomes of the A genome might provide a powerful tool for identifying the A genome donor.

Analysis with a single highly repeated DNA sequence

A limitation in the use of total highly repeated DNA is that it consists of a population of different sequences so that significant changes in component sequences in different wheat species might not necessarily be disclosed by the hybridization analyses. If, on the other hand, the radioactive probe used for chromosome mapping was a single, pure, highly repeated sequence, then changes in chromosomal patterns would be readily detectable. A buoyant density satellite DNA provided such a probe. When the total nuclear DNA of Chinese Spring wheat is centrifuged in Cs_2SO_4 gradients containing Ag^+, a small proportion of the DNA is separated from the remainder of the nuclear DNA because its density is increased by preferential binding of Ag^+ (Figure 3.1). Reassociation kinetics, restriction enzyme analysis and sequencing of complementary RNA transcripts of this DNA showed it to be composed of tandem repeats of a simple sequence (Dennis *et al.*, 1980). Direct sequencing of cloned segments of this satellite confirmed this conclusion, showing the sequence to be composed of combinations of the triplets $\frac{GAA}{CTT}$ and $\frac{GAG}{CTC}$ (Figure 3.2).

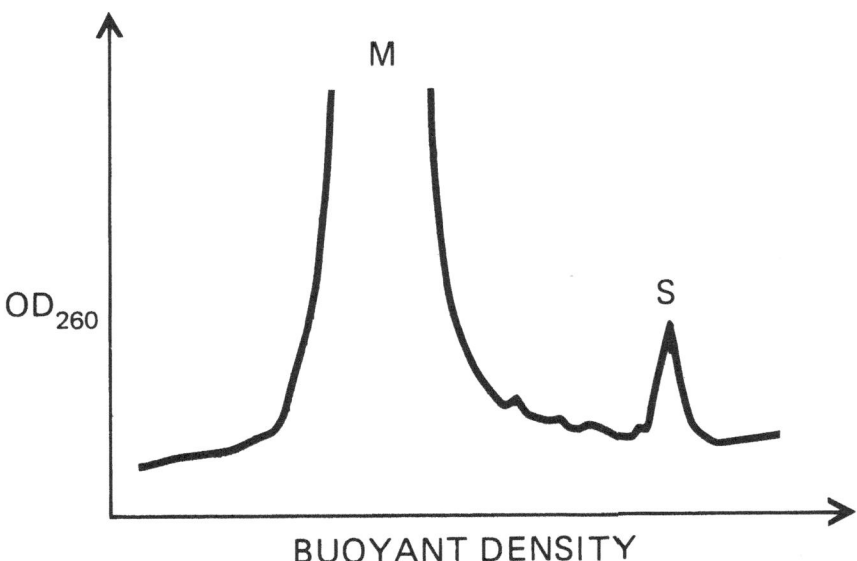

Fig. 3.1. Isolation of a satellite DNA from hexaploid wheat. Total DNA from wheat (Chinese Spring) was centrifuged to equilibrium in a Ag^+/Cs_2SO_4 gradient. A satellite DNA (S) separates from the mainband DNA (M).

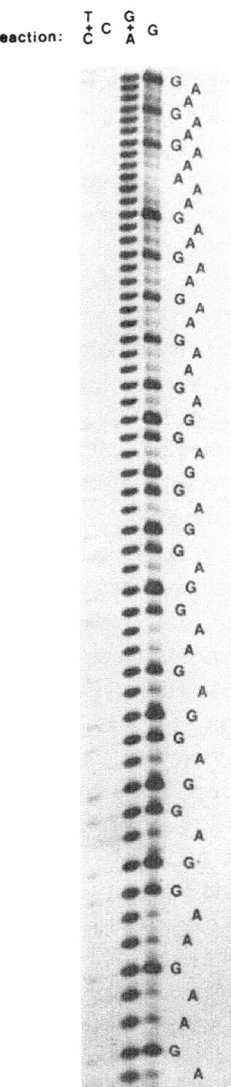

Fig. 3.2. Nucleotide sequence of a cloned segment of Ag⁺-satellite DNA from wheat. The base sequence was determined chemically using rapid sequencing techniques (Maxam and Gilbert, 1977). A resultant autoradiograph is shown with the deduced base sequence of the satellite DNA.

In situ hybridization with this sequence gave precisely the same pattern as did the total highly repeated DNA (Figure 3.3). All seven chromosomes of the B genome and chromosomes 4 and 7 of the A genome contained numbers of major sites, with some other chromosomes of the A and D genomes having minor sites (Figure 3.3a). The identity of each chromosome was checked by *in situ* hybridization to the ditelocentric marker chromosomes generated by Sears (1954). The major sites of tandem repeats of the triplet sequences are in centromeric, interstitial and terminal locations. This enables specific identification of many of the chromosome arms, so that even if introgression of chromosomes from different genomes or rearrangements of segments of chromosomes have occurred, we have some chance of recognising these events. The congruence of the sites of the Ag^+ satellite and total repeated DNA in hexaploid and tetraploid wheats implies that since the time of origin of hexaploid wheats there has been no change in the representation or distribution of highly repeated sequence DNA in the chromosome complements. Archaeological evidence places the time of emergence of hexaploids to be about 10 000 years ago (Harlan, Chapter 1). In addition, the primitive and cultivated tetraploids have identical patterns.

In situ hybridization has identified a distinction between the Russian (*T. timopheevi*) and Mediterranean tetraploids. The presumptive chromosome 4B of *T. timopheevi* differs from its counterpart in the Mediterranean wheats by a possible pericentric rearrangement as detected by a shift in the location of a major site of the Ag^+-satellite sequence (Gerlach *et al.*, 1978). The A genome of the Russian species contains a chromosome with a pattern that is not represented in the Mediterranean tetraploids. Historically it has been assumed that the B genome ancestors were not the same for these two groups of wheat species. Our observations suggest that this may not have been so, since the B genome patterns differ by only a single site. The A genomes show more substantial differences.

We have argued that because the basic pattern of the major sites has remained constant in the polyploid wheats, the Ag^+-satellite may provide a way of identifying the donors of both the A and B genomes among wild diploid wheats. The donor of the B genome would be expected to have these sequences accounting for approximately 3% of its DNA and to have major locations of the sequences on all chromosomes. *T. urartu* does not fulfil either of these criteria and can be definitely ruled out. In common with *T. monococcum* and *T. boeoticum*, *T. urartu* has only a small amount of the sequence and does not contain any major chromosomal locations (Figure 3.4). On the other hand, there are several *Aegilops* species which have a significant concentration of these satellite sequences on every chromosome of the complement (Figure 3.5). All species are in the Sitopsis section of the genus. *Ae. longissima* most closely resembles the B genome in its pattern of sites. Chromosomes 2B, 5B and 7B have direct

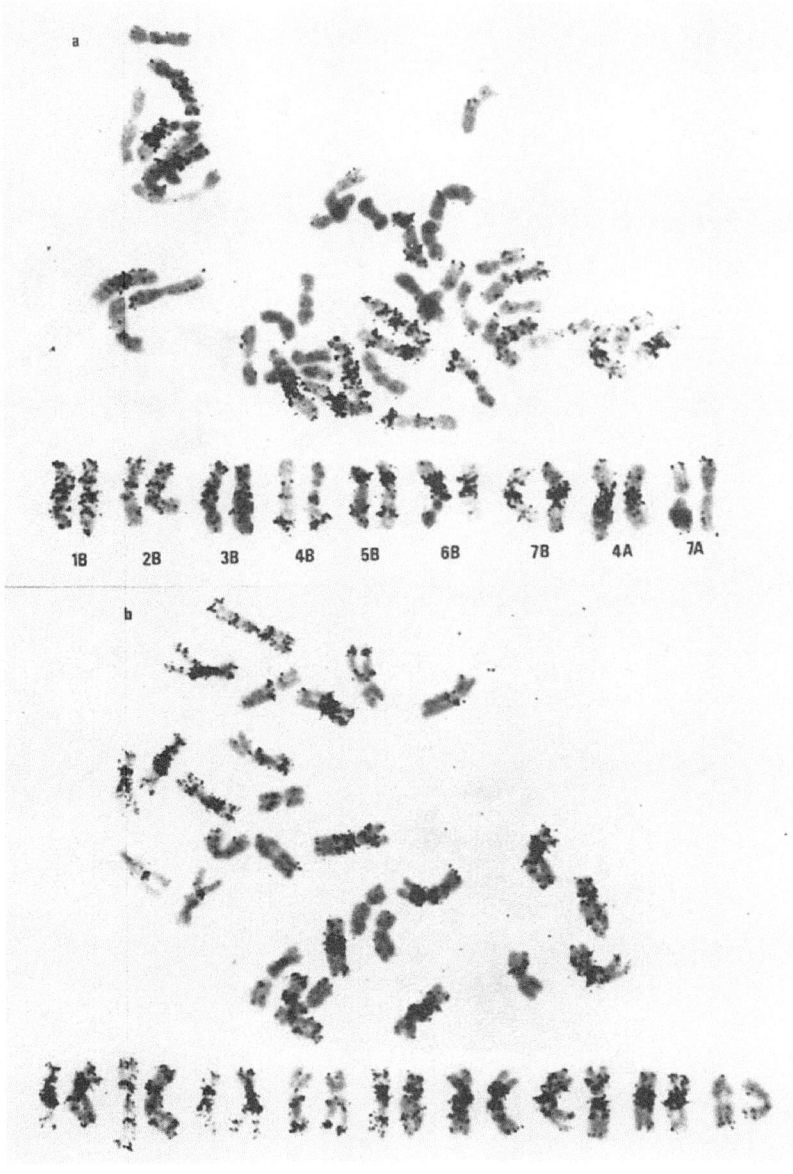

Fig. 3.3. Chromosome localisation of the Ag[+]-satellite DNA from wheat. *In situ* hybridization of [3]HcRNA to metaphase chromosomes from:
(a)   hexaploid *T. aestivum* cv. Chinese Spring; and
(b)   tetraploid *T. dicoccoides*.
Identities of labelled chromosomes are shown.

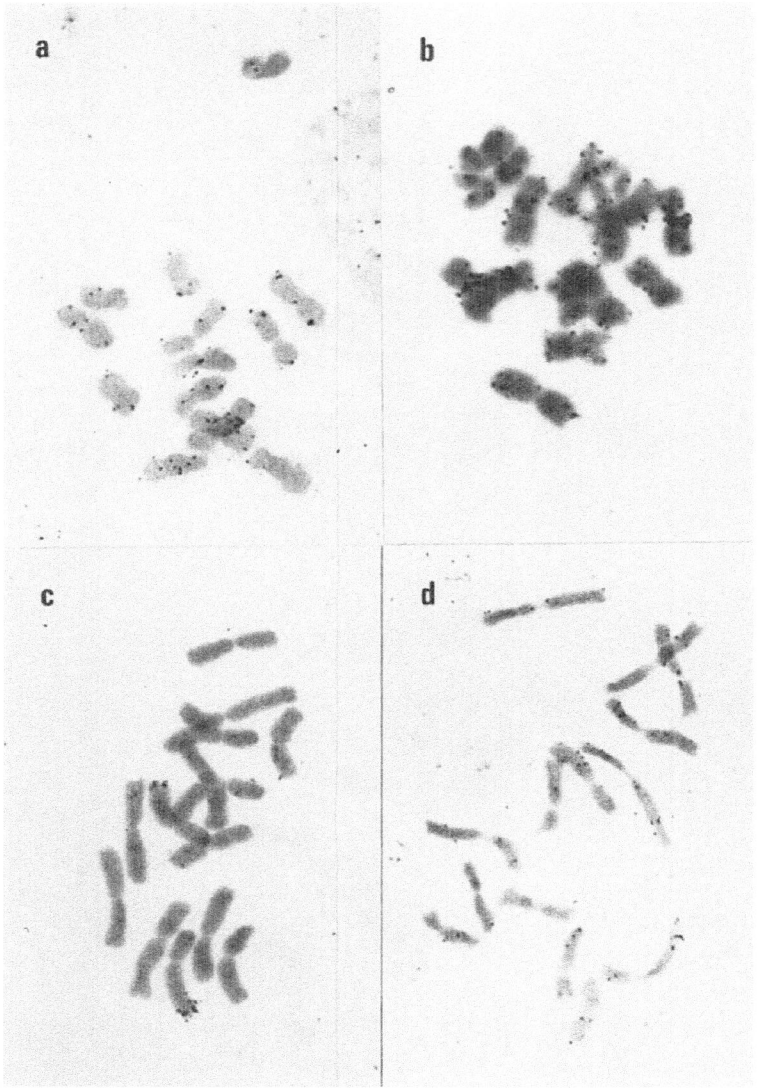

Fig. 3.4. Chromosomal localisation of the Ag^{+}-satellite DNA of wheat by *in situ* hybridization to root tip metaphase chromosomes in the diploid species:
(a)   *T. urartu*
(b)   *T. monococcum*
(c)   *T. boeoticum*
(d)   *Ae. squarrosa*

counterparts in *Ae. longissima* and 1B is also similar. *Ae. speltoides*, long favoured as the B genome donor (Sarkar and Stebbins, 1956; Riley *et al.*, 1958), does have major sites on all chromosomes but the pattern does not resemble that found in the polyploids. A species collected only recently, *Ae. searsii*, has been proposed as the donor of the B genome (Feldman, 1978), but its pattern of sites also fails to correspond to that of the B genome of the modern species.

None of the proposed B genome donors has a pattern of satellite sites identical to that in the polyploids. It is possible that other wild species will show greater correspondence to the B genome pattern. Since a characteristic feature of highly repeated DNA is that changes of repetition frequency can occur between species (Peacock *et al.*, 1980) it is also possible that alterations of the pattern of sites have occurred since the formation of the tetraploid wheats but the stability of pattern throughout the range of primitive and cultivated polyploids makes this unlikely. It is clear that the donor(s) of the B genome must be closely related to the species in the Sitopsis section of *Aegilops*.

The Ag^+-satellite probe also provides information relating to the origin of the A genome. One accession of *T. boeoticum* has the small telomeric block of sequences on one chromosome similar to that found on chromosome 7A of some polyploid wheats (Figure 3.4). Neither *T. boeoticum* nor any other of the diploid species contains a chromosome with the distinctive pattern of sites characteristic of chromosome 4A. Perhaps this is not surprising since other data suggest that 4A may have had a complex evolutionary history. In crosses of hexaploid wheat with *T. boeoticum*, chromosome 4A pairs with a *boeoticum* homoeologue (Chapman and Riley, 1966), but when the hexaploid wheat is crossed to *T. urartu* it is the only chromosome of the A genome which fails to pair with a chromosome from the diploid species (Chapman *et al.*, 1976; Dvorak, 1976). Furthermore, in substitution analysis chromosome 4A can be replaced by a chromosome from *Ae. sharonensis* (Miller and Chapman, pers. comm.), a species from the Sitopsis section and one of the contenders for B genome ancestry.

Can the ribosomal RNA genes define the ancestry of modern wheat?

In higher organisms, the genes coding for the major ribosomal RNAs are reiterated and organised in tandem arrays. The repeating unit contains the coding sequences for the 18S and 26S RNAs together with spacer sequences. Characteristically the coding sequences are conserved but the sequences within spacer regions can show considerable diversity between species. These genes, therefore, provide another possible tool for the identification of the diploid progenitors of polyploid wheats.

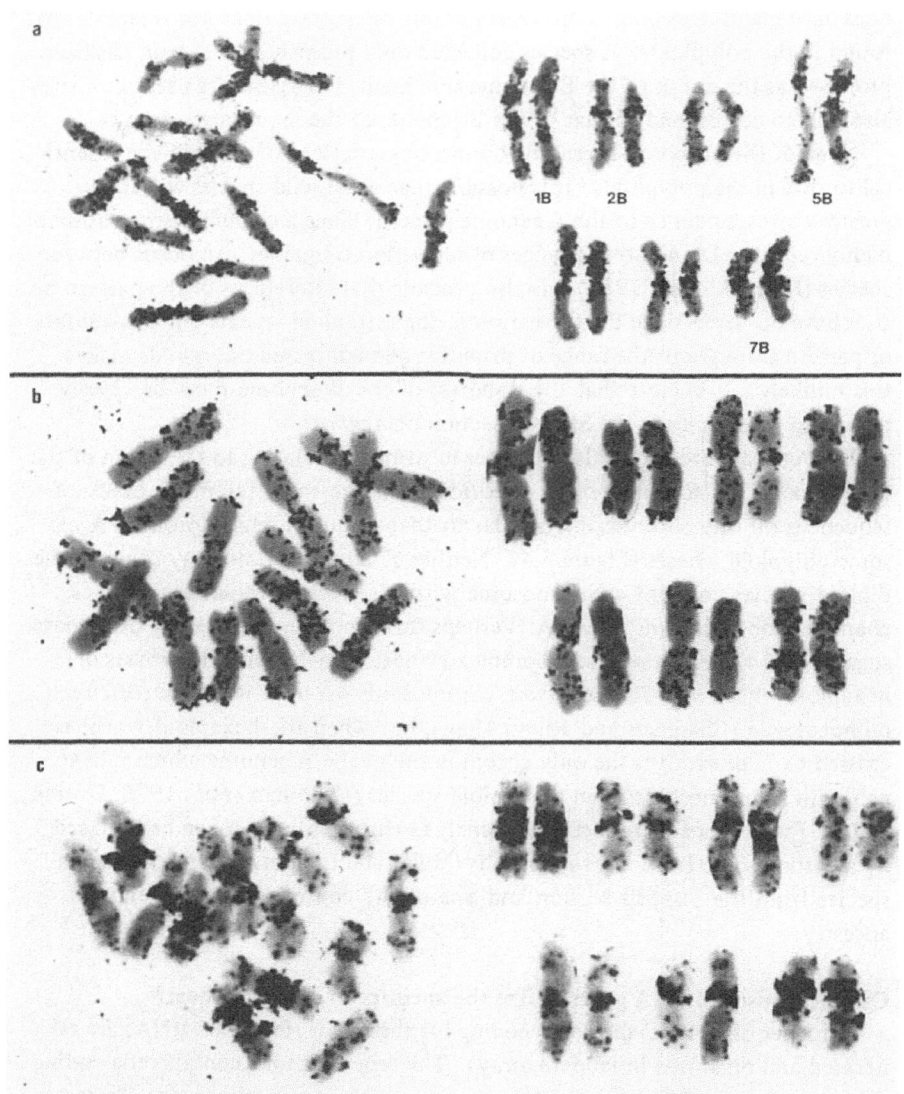

Fig. 3.5. Chromosomal localisation of the wheat Ag[+]-satellite in species in the Sitopsis section of *Aegilops*. Metaphase spreads and karyotypes are shown for each species. Chromosomes of *Ae. longissima* which have patterns similar to B genome chromosomes from polyploid wheat are marked: (a) *Ae. longissima*, (b) *Ae. searsii*, and (c) *Ae. speltoides*.

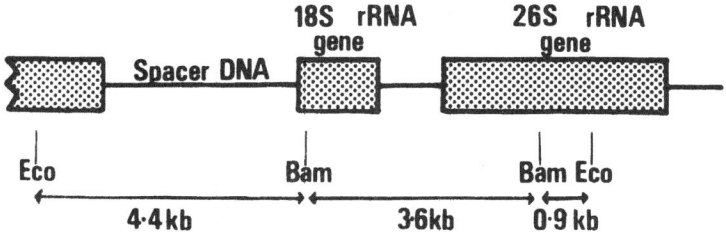

Fig. 3.6. Diagram of the ribosomal RNA gene repeating unit from wheat. Sites for the restriction enzymes BamH1 and EcoR1 are indicated. The fragment lengths obtained following simultaneous digestion with these two enzymes are shown.

In hexaploid wheat (cv. Chinese Spring) there are approximately 2300 copies of the repeating ribosomal gene unit per haploid genome (Appels *et al.*, 1980). The repeating unit is 9 kb and contains both EcoR1 and BamH1 restriction enzyme recognition sites (Figure 3.6). *In situ* hybridization has shown that in the cultivar Chinese Spring, 90% of the ribosomal RNA genes are on chromosomes 1B and 6B with the remaining repeat units being located on chromosome 5D (Figure 3.7). This is a surprising result since we would have expected ribosomal RNA genes to be present on each of the component genomes of the hexaploid. For example, all diploid wheat species being considered as A genome donors have two chromosomal sites for the ribosomal RNA genes (Gerlach *et al.*, 1980). This argues that the genome of the hexaploid wheat Chinese Spring cannot be ascribed simply to two successive hybridization events, without any changes in the chromosomes. Either some A genome chromosomes or segments of the chromosomes carrying the ribosomal genes have been lost in chromosome rearrangements or introgression, or else there has been a diminution of the total numbers of ribosomal gene repeats, to such an extent on the A genome chromosomes that they are no longer detectable. The numbers of ribosomal repeats in hexaploid, tetraploid and diploid wheat species vary, but are of a comparable order of magnitude (Mohan and Flavell, 1974; Flavell and Smith, 1974; Liang *et al.*, 1978). The B genome sites are preferentially retained in the hexaploid.

The majority of repeat units have identical restriction maps, and since they are primarily in two B chromosome sites they could provide another means of identifying the B genome ancestor. We have therefore looked at the molecular organisation of the ribosomal genes. EcoR1 and BamH1 digests of several diploid and polyploid species are shown in Figure 3.8.

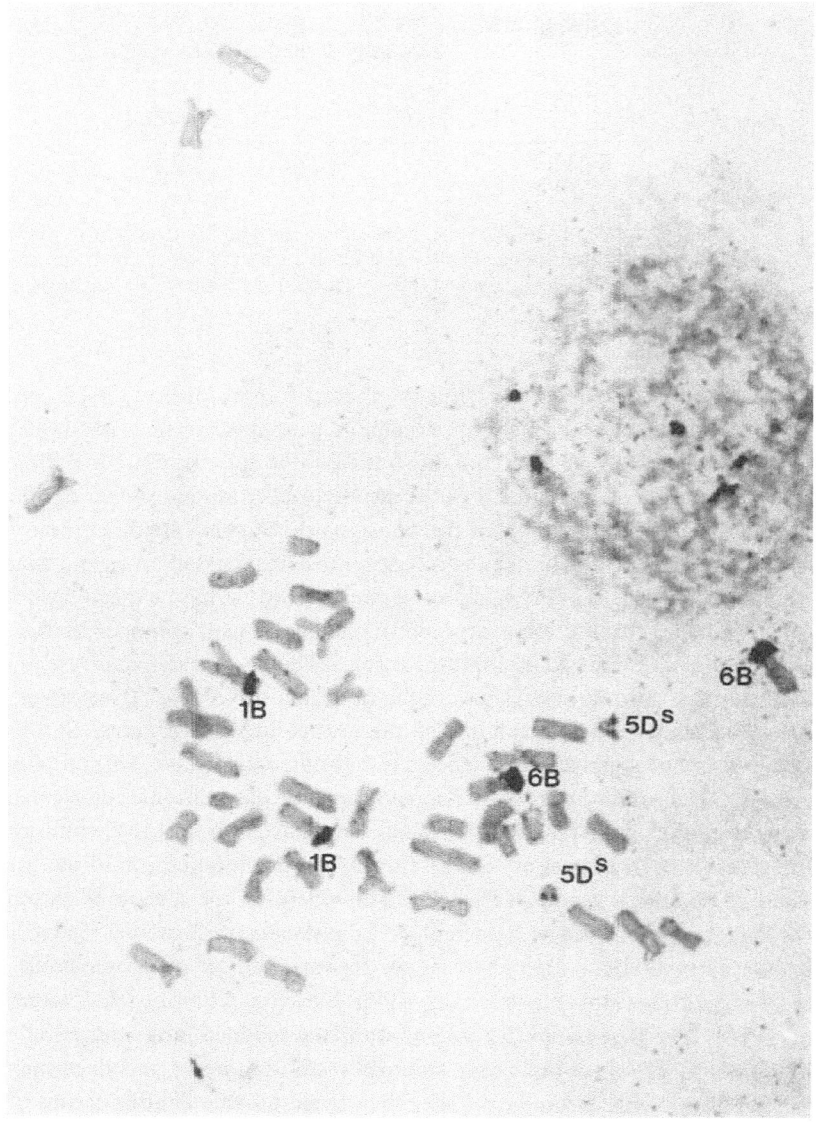

Fig. 3.7. Localisation of the ribosomal RNA genes in hexaploid wheat *T. aestivum* cv. Chinese Spring. *In situ* hybridization of ^{125}I-18S and 26S RNA to chromosomes of a 5D double ditelocentric tester stock. Chromosomes containing ribosomal RNA sites are indicated. The sites are also visible in the early prophase nucleus.

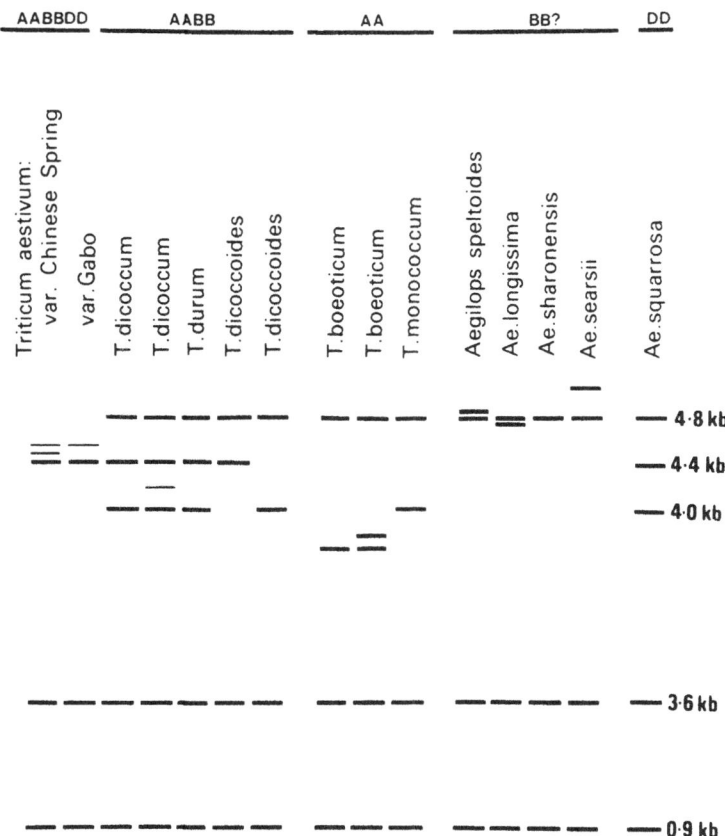

Fig. 3.8. Ribosomal RNA gene fragment lengths in a number of wheat species. DNA was digested with restriction enzymes EcoR1 and BamH1, separated according to size by electrophoresis and hybridized with radioactive probe for the ribosomal RNA repeating unit. In some species two accessions were examined.

In Chinese Spring, the restriction enzymes produce a characteristic pattern of molecular fragments. Two segments, 0.9 and 3.6 kb long, are derived from the 18S and 26S coding regions and a 4.4 kb segment primarily from the spacer region (Figure 3.6). There are two other minor variants of this spacer segment, the molecules being 4.5 and 4.6 kb long respectively. Another hexaploid wheat, *Gabo*, has precisely the same coding region segments and the same major spacer segment but lacks the shorter minor spacer variant (Figure 3.8). The tetraploid wheats show a markedly different pattern for the spacer region. Two accessions of the primitive tetraploid *T. dicoccoides* have different spacer bands. Neither of these is identical to the patterns in two accessions of the cultivated tetraploid, *T. dicoccum*. Another tetraploid, *T. durum*, has a pattern identical to one of the *T. dicoccum* patterns. The spacer variability does not necessarily mean that each polyploid has had a unique origin, since localised amplification events can occur in tandem arrays of repeating sequences resulting in a particular molecular variant being expanded.

One spacer length that occurs in all polyploids, other than one accession of *T. dicoccoides*, is the 4.4 kb segment, which we know to be located on chromosome 1B and 6B in Chinese Spring (Appels *et al.*, 1980). We might expect A genome donors not to have this 4.4 kb segment and this is the case. They do contain a 4.8 kb segment comparable to that seen in the tetraploid. Only *T. monococcum* has a 4.0 kb segment. When the presumptive B genome donors are examined, no species contains the 4.4 kb band! Instead they all contain the 4.8 kb band and, in addition, each species has its own characteristic set of other fragment lengths. Another perplexing result is that the only diploid species to show a band of length 4.4 kb is *Ae. squarrosa* which is regarded as a DD diploid. *Ae. squarrosa* also contains both of the 4.8 kb and 4.0 kb bands present in most tetraploid wheats. Does this mean that the ribosomal RNA genes from *Ae. squarrosa* have become incorporated into the B genome of polyploid wheats?

5S ribosomal RNA genes as probes

The third probe we have used for investigating the phylogeny of wheats is the 5S ribosomal RNA gene. The units coding for the 5S RNA genes of Chinese Spring are of two different sizes, each organised in tandem arrays (Appels *et al.*, 1980).

The restriction enzyme BamH1 defines two fragments of length 420 bp and 500 bp which hybridizes radioactive 5S RNA probe. Each segment contains one 5S RNA gene, 120 bp long, and spacer DNA. The tandem arrays of the 420 bp repeating unit have a major location on chromosome 1B, so this 420 bp segment could be definitive for this particular chromosome region.

We have examined a range of wheats for the presence of the 420 bp unit (Figure 3.9). The hexaploid and tetraploid wheats are identical, containing both 500 and 420 bp length fragments. This demonstrates stability in this character since the formation of tetraploid wheats, and encouraged us to probe the diploid species. The three AA diploids examined do not contain the 420 bp segment.

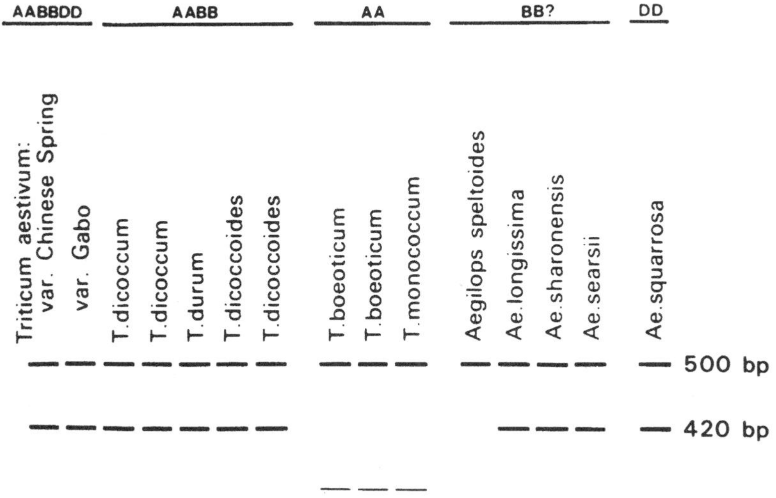

Fig. 3.9. The lengths of the 5S ribosomal RNA gene repeats in a number of hexa-ploid, tetraploid and diploid wheat and *Aegilops* species. DNA was digested with restriction enzyme BamH1, separated according to size by electrophoresis and hybridized with radioactive probe for the 5S RNA genes. In some species more than one accession was examined.

This is consistent with the 420 bp gene having its location in the B genome. *Ae. speltoides* has only the 500 bp repeat. Other *Aegilops* species, *Ae. longissima*, *sharonensis* and *searsii*, all contain the 420 bp repeat unit in addition to the 500 bp repeat, consistent with them being closely related to the source of the B genome. *Ae. squarrosa* contains both the 420 bp and 500 bp repeat unit.

The 420 bp repeat in the diploids could not be differentiated from the 420 bp repeat in the polyploid wheats by restriction enzyme analysis. A particular HaeIII recognition site is present in all 420 bp repeats.

Discussion

The three repeated DNA probes, the Ag$^+$-satellite DNA, the ribosomal RNA and 5S RNA gene repeating units have all yielded information on the relationship of the genomes of the modern polyploid wheats and the genomes of wild diploid grass species. The Ag$^+$-satellite has its major concentration in the B genome and we found that several *Aegilops* species had comparable contents of this highly repeated sequence, thus giving support to the view that the B genome donor might well be one of the wild diploid *Aegilops* species. The pattern of distribution of the Ag^+-satellite DNA sequences in hexaploid wheat was also consistent with the D genome being derived from *Ae. squarrosa* — neither complement of chromosomes contained any major sites of the sequence (Figures 3.3,3.4).

The diploid *Triticum* species which have been proposed as likely progenitors for the A genome also lack major sites for this sequence and differ from the polyploids' A genome. In both tetraploid and hexaploid wheats chromosome 4A has heavy concentrations of the sequence and chromosome 7A has terminal sites on one or both arms. By this criterion none of the diploid *Triticum* species can be identified as the participant in a hybridization event with an *Aegilops* species in the formation of tetraploid wheat. The pattern and content of the sequences in chromosome 4A are suggestive of an origin for this chromosome different from that of the remaining six chromosomes of the A genome. Chromosome 4A may have been derived by introgression from a species more closely related to the diploids of the Sitopsis section of *Aegilops*.

The Ag^+-satellite DNA was also important in demonstrating the close relatedness of the tetraploid and hexaploid wheats and supports the accepted notions of their successive formation. The patterns of distribution on the seven chromosomes of the B genome and chromosomes 4A and 7A were identical in a number of hexaploids and tetraploids we examined. There were minor differences at the tetraploid level between *T. timopheevi* and *T. dicoccum*. Another minor difference, the loss of one site on an arm of 7A, occurs within the polyploid wheats. This latter difference is explicable in terms of modulation of the number of repeats, a property well documented for highly repeated DNA (Peacock *et al.*, 1980).

The ribosomal RNA genes were also largely contained in the B genome of Chinese Spring. There must have been selective loss of the ribosomal genes from the A and D genomes. The plasticity of this particular component of the genome is also shown by the dramatic differences in their spacer DNA restriction patterns. This applies to the hexaploids and tetraploids which were so similar when examined by the Ag^+-satellite probe. Unfortunately, we lack *in situ* hybridization data for tetraploid wheats. Restriction enzyme analyses show most of them to contain DNA segments characteristic of the diploid *Triticum* species thought to be related to the A genome, whereas the hexaploids we have examined do not have any representatives of this particular segment class. Thus, tetraploid wheats should also have A genome locations for these genes.

The 5S RNA genes have one major site on chromosome 1B and the 420 bp repeat unit is largely restricted to this site (Appels *et al.*, 1980). The 500 bp unit length repeats occur in other sites in the genome which have not been mapped.

All hexaploids and tetraploid species have these two different repeat units. So too do all of the diploid *Aegilops* species, other than *Ae. speltoides.* The proposed D genome donor, *Ae. squarrosa* has approximately equal numbers of the two units even though in the hexaploid wheat the 420 bp unit is certainly restricted to chromosome 1B and is not detectable on any of the D genome chromosomes.

Conclusions

With these three repeated DNA probes, although they represent a minor proportion of the total population of different repeated DNAs in the wheat genome, we have been able to come to some conclusions about the A, B and D genomes of wheat and place some restrictions on their potential origins. Previously suggested donors of the A genome have included *T. boeoticum* and *T. monococcum*. However, the patterns we have observed with the Ag^+-satellite probe excludes these proposals in their simplest form. Either there have been events more complex than a straightforward hybridization, or the appropriate diploid species has not yet been identified. The ribosomal RNA probe also excludes any simple equation of the diploid *Triticum* species and the A genome of the polyploid wheats. The restriction enzyme generated bands of the spacer regions in the diploid species are not present at all in some hexaploids and are only partially represented among tetraploids. The 5S RNA probe consolidates this argument since both *T. boeoticum* and *T. monococcum* have a restriction band which is not found in any of the polyploid wheats.

The Ag^+-satellite pattern has confirmed that the diploid ancestor of the B genome almost certainly is an *Aegilops* species of the Sitopsis section of that genus. None of the previously favoured diploid species has precisely the same pattern of sites as the polyploid wheats. The diploid species with greatest similarity to the polyploid B genome pattern is *Ae. longissima*. However, this species does not have the same ribosomal RNA gene organisation as any of the B genomes of the polyploid wheats. Nor, for that matter, do any of the other diploid species. The 5S gene patterns exclude *Ae. speltoides* but are consistent with the other species of the Sitopsis section being closely related to the B genome.

The D genome presents a puzzling situation. The Ag^+-satellite result is consistent with *Ae. squarrosa* as the D genome donor. The 5S RNA probe shows an identity of the *Ae. squarrosa* pattern with the B genome pattern. The 420 bp repeat, which represents approximately 50% of the 5S genes in *Ae. squarrosa*, is not detectable in the hexaploid wheat D genome. The ribosomal RNA gene probe also poses problems for the *Ae. squarrosa*-D genome equivalence, because we were able to detect only small numbers of ribosomal genes on the D genome. Furthermore, *Ae. squarrosa* is the only diploid species which contains a ribosomal segment which is characteristic of the B genome of both tetraploid and hexaploid wheats. *Ae. squarrosa* may have had an involvement with both the B and D genomes of polyploid wheats!

References

Appels, R., Gerlach, W.L., Dennis, E.S., Swift, H. and Peacock, W.J. (1980) Molecular and chromosomal organization of DNA sequences coding for the ribosomal RNAs of cereals. *Chromosoma* **78**, 293–311.

Chapman, V. and Riley, R. (1966) The allocation of the chromosomes of *Triticum aestivum* to the A and B genomes and evidence of genome structure. *Can. J. Genet. Cytol.* **8**, 57–63.

Chapman, V., Miller, T.E. and Riley, R. (1976) Equivalence of the A genome of bread wheat and that of *Triticum uartu. Genet. Res.* **27**, 69–76.

Dennis, E.S., Gerlach, W.L. and Peacock, W.J. (1980) Identical polypyrimidine-polypurine satellite DNA in wheat and barley. *Heredity* **44**, 345–366.

Dhaliwal, H.S. (1976) Fertility and morphology of synthetic amphiploids and the origin of tetraploid wheats. *Cereal Res. Comm.* **4**, 411–418.

Dhaliwal, H.S. and Johnson, B.L. (1976) Anther morphology and the origin of the tetraploid wheats. *Am. J. Bot.* **63**, 363–368.

Dover, G.A. and Riley, R. (1972) Prevention of pairing of homoeologous meiotic chromosomes of wheat by an activity of supernumerary chromosomes of *Aegilops. Nature* **240**, 159–161.

Dvorak, J. (1976) The relationship between the genome of *Triticum urartu* and the A and B genomes of *Triticum aestivum. Can. J. Genet. Cytol.* **18**, 371–377.

Feldman, M. (1978) New evidence on the origin of the B genome of wheat. Proc. 5th Int. Wheat Genetics Symp., New Delhi, 1978. Vol. **1**, pp. 120–132.

Flavell, R.B. and Smith, D.B. (1974) Variation in nucleolar organizer rRNA gene multiplicity in wheat and rye. *Chromosoma* **47**, 327–334.

Flavell, R.B., Rimpau, J. and Smith, D.B. (1977) Repeated sequence DNA relationships in four cereal genomes. *Chromosoma* **63**, 205–222.

Flavell, R.B., O'Dell, M. and Smith, D.B. (1979) Repeated sequence DNA comparisons between *Triticum* and *Aegilops* species. *Heredity* **42**, 309–322.

Gerlach, W.L. and Peacock, W.J. (1980) Chromosomal locations of highly repeated DNA sequences in wheat. *Heredity* **44**, 269–276.

Gerlach, W.L., Appels, R., Dennis, E.S. and Peacock, W.J. (1978) Evolution and analysis of wheat genomes using highly repeated DNA sequences. Proc. 5th Int. Wheat Genetics Symp., New Delhi, 1978. Vol. **1**, pp. 81–91.

Gerlach, W.L., Miller, T.E. and Flavell, R.B. (1980) The nucleolus organizers of diploid wheat revealed by *in situ* hybridization. *Theor. Appl. Genet.* (in press).

Jaaska, V. (1980) Electrophoretic survey of seedling esterases in wheats in relation to their phylogeny. *Theor. Appl. Genet.* **56**, 273–284.

Johnson, B.L. (1975) Identification of the apparent B-genome donor of wheat. *Can. J. Genet. Cytol.* **17**, 21–39.

Johnson, B.L. and Dhaliwal, H.S. (1978) *Triticum urartu* and genome evolution in the tetraploid wheats. *Am. J. Bot.* **65**, 907–918.

Liang, G.H., Wang, A.S. and Phillips, R.L. (1978) Control of ribosomal RNA gene multiplicity in wheat. *Can. J. Genet. Cytol.* **19**, 425–435.

Maxam, A. and Gilbert, W. (1977) A new method for sequencing DNA. *Proc. Nat. Acad. Sci. U.S.A.* **74**, 560–564.

McFadden, E.S. and Sears, E.R. (1946) The origin of *Triticum spelta* and its free threshing hexaploid relatives. *J. Hered.* **37**, 81–89, 107–116.

Mohan, J. and Flavell, R.B. (1974) Ribosomal RNA cistron multiplicity and nucleolar organizers in hexaploid wheat. *Genetics* **76**, 33–44.

Natarajan, A.T. and Sharma, N.P. (1974) Chromosome banding patterns and the origin of the B genome in wheat. *Genet. Res.* **21**, 103–8.

Nishikawa, K. and Furuta, Y. (1978) DNA content of nucleus and individual chromosomes and its evolutionary significance. Proc. 5th Int. Wheat Genetics Symp., New Delhi, 1978. Vol. 1, pp. 133–140.

Peacock, W.J., Dennis, E.S. and Gerlach, W.L. (1980) Satellite DNA — change and stability. *Chromosomes Today* **7** (in press).

Rees, H. and Walters, M.R. (1965) Nuclear DNA and the evolution of wheat. *Heredity* **20**, 73–82.

Riley, R. and Chapman, V. (1958) Genetic control of the cytologically diploid behaviour of hexaploid wheat. *Nature* **182**, 713–715.

Riley, R., Unrau, J. and Chapman, V. (1958) Evidence on the origin of the B genome of wheat. *J. Heredity* **49**, 91–98.

Sarkar, P. and Stebbins, G.L. (1956) Morphological evidence concerning the origin of the B genome in wheat. *Am. J. Bot.* **42**, 297–304.

Sears, E.R. (1954) The aneuploids of common wheat. *Mo. Agric. Exp. Stn. Res. Bull.* **572**, 1–58.

Sears, E.R. (1956) The B genome of *Triticum. Wheat Inf. Serv.* **4**, 8–10.

Sears, E.R. and Okamoto, M. (1958) Intergenomic relationships in hexaploid wheat. *Proc. 10th Int. Congr. Genet.* **2**, 258–259.

Suemoto, H. (1978) The origin of the cytoplasm of tetraploid wheat — III. Proc. 5th Int. Wheat Genetics Symp., New Delhi, 1978. Vol. 1, pp. 273–281.

Vitozzi, L. and Silano, V. (1976) The phylogenesis of protein α-amylase inhibitors from wheat seed and the speciation of polyploid wheat. *Theor. Appl. Genet.* **48**, 279–284.

Zohary, D. and Feldman, M. (1962) Hybridization between amphidiploids and the evolution of polyploids in the wheat (*Aegilops-Triticum*) group. *Evolution* **16**, 44–61.

4 MOLECULAR ASPECTS OF WHEAT EVOLUTION: RUBISCO COMPOSITION

S.G. Wildman

Background

In 1975, Chen, Gray and Wildman analyzed the composition of rubisco* in an endeavour to decide what diploid species among the *Triticinae* would have been eligible to serve as female partners in the evolution of tetraploid and hexaploid wheats. Rubisco is a very large protein weighing about 550 000 daltons and is the enzyme that catalyzes the combination of atmospheric carbon dioxide with ribulose-1,5-diphosphate during photosynthesis. Rubisco often accounts for one-quarter of the total protein contained in Angiosperm leaves. Its presence in such large quantity makes it relatively simple to obtain rubisco in pure form for analysis of its composition.

The rubisco quaternary structure is constructed from eight large subunits combined with eight small subunits. When S-carboxymethylated rubisco is electrofocussed in the presence of 8M urea, the large and small subunits dissociate and the monomeric subunits resolve into individual polypeptides. The large subunit resolves into a cluster of three polypeptides which differ slightly in isoelectric points. In the case of rubisco from *Aegilops squarrosa*, three polypeptides with isoelectric points in the neighbourhood of 6.2, 6.3, and 6.4 are resolved. The *Ae. squarrosa* large subunit can be distinguished from the large subunit of *Ae. speltoides* rubisco whose three polypeptides have relative isoelectric points of 6.0, 6.1, and 6.2. Thus, two out of the cluster of three *Ae. speltoides* large subunit polypeptides are different from the *Ae. squarrosa* cluster of three polypeptides. Genetic analyses employing electrofocussing behaviour of rubisco isolated from reciprocal, inter-specific hybrids, *Triticum boeoticum* x *T. dicoccoides* being only one of several examples (Sakano *et al.*, 1974; Chen *et al.*, 1975a), have shown the isoelectric points of the cluster of three polypeptides to behave as a genetic unit. Genetic information controlling the isoelectric points of the cluster is inherited exclusively via the maternal line and is therefore contained in extranuclear DNA. It is suspected that the three polypeptides constituting a cluster are of post-transcriptional origin from a single gene (Gray

*An acronym coined by Prof. David Eisenberg to substitute for 'Fraction 1 protein', 'carboxydismutase', 'ribulose diphosphate carboxylase', 'ribulose-1,5-bisphosphate carboxylase-oxygenase', etc.

et al., 1978). Restriction endonuclease analysis has pinpointed chloroplast DNA as possessing the genetic information controlling the sequence of amino acids in the rubisco large subunit (Coen *et al.*, 1977).

The results obtained by Chen *et al.* for rubisco of wheat species are displayed in diagrammatic form in Figure 4.1. Based on an exclusive maternal inheritance of the genetic information coding for the rubisco large subunit, they concluded that a female parent contributed the BB genome in the cross that produced tetraploid *T. dicoccum*, the latter species in turn serving as female parent to perpetuate the BB genome in combination with the DD genome provided by the pollen of *Ae. squarrosa* in the evolution of hexaploid wheat, *T. aestivum*. Concerning the question of whether *T. urartu* or *Ae. speltoides* was the likely female donor of the BB genome, the rubisco large subunit evidence appeared to eliminate *T. urartu* as an immediate progenitor of *T. dicoccum*. Reichenbächer *et al.* (1977), using immunoelectrophoresis of undenatured rubisco, have confirmed the basic findings of Chen *et al.* Vedel *et al.* (1978) isolated chloroplast DNA from species in the *Triticinae* and compared electrophoregrams of DNA fragments after EcoRI cleavage and arrived at the same conclusion. They also isolated and cleaved mitochondrial DNA's and reached the additional conclusion that the BB genome of wheat could not have originated from the contemporary form of *Ae. speltoides*.

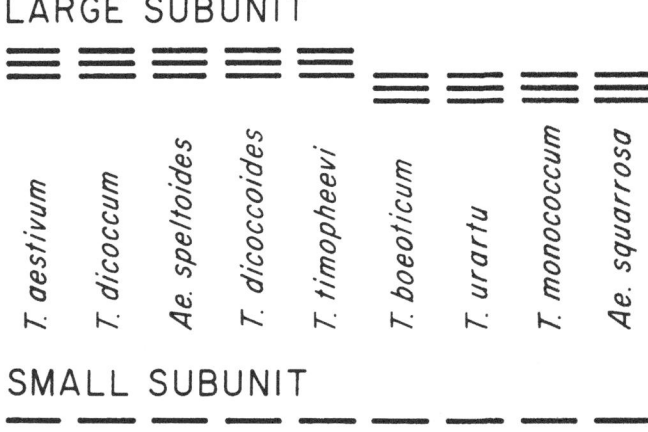

Fig. 4.1. Diagrammatic representation of the polypeptide composition of rubisco contained in species of the *Triticinae* as resolved by electrofocussing. Adapted from Chen *et al.* (1975a).

Whereas the composition of the rubisco large subunit served as a useful indicator for aspects of wheat evolution, no help was obtained from comparative analysis of the small subunit since only a single polypeptide of the same iso-electric point appeared in all of the rubiscos analysed as also shown in Figure 4.1. This is a rather striking condition when compared to rubisco contained in Dicotyledons. Analysis of more than 175 species of plants representing several phyla and numerous genera of the Plant Kingdom showed the rubisco small subunit to be more often heterogeneous in polypeptide composition (Chen and Wildman, 1980a) than homogeneous as in the wheat species. However, absence of isoelectric point differences does not necessarily mean that there are no differences in sequence arrangement of amino acids, or that there are no differences in amino acid composition. In fact, I believe that a thorough study of the composition and sequence of the small subunit of rubisco obtained from different accessions of the wheat species might cast additional light on wheat phylogeny. But to develop this view, it is necessary to digress at this point to examine the course of rubisco evolution among polyploid species of Dicotyledons.

Evolution of rubisco in Dicotyledons

Among 65 species of *Nicotiana*, 29 different species of rubisco are now known which display different electrofocussing compositions (Chen and Wildman, 1980a). Figure 4.2 illustrates some of the different kinds of *Nicotiana* rubisco. Evolution of the large subunit has occurred to the extent that four types of polypeptide clusters can be recognized in today's *Nicotianas* compared with the two types found in rubisco of *Triticinae* species. In contrast to the absence of differences in the small subunit of wheat species, the small subunit of *Nicotiana* rubisco has evolved so that 13 polypeptides of different isoeletric points are now present among the 29 species of rubisco. Furthermore, the small subunit polypeptides of different isoelectric points have been combined in different ways during evolution of *Nicotiana* species. As also shown in Figure 4.2, the rubisco small subunit of one *Nicotiana* species may consist of only a single polypeptide (as in wheat species), or the small subunit may consist of as many as four different kinds of polypeptides.

Origin of *Nicotiana* rubisco with two or more kinds of small subunit polypeptides by amphiploidy

The small subunit of tobacco (*Nicotiana tabacum*) rubisco is composed of polypeptides #7 and #9 as indicated in Figure 4.2. Goodspeed (1954) showed that *N. tabacum* (n = 24) arose by amphidiploidy following interspecific hybridization between *N. tomentosiformis* (n = 12) and *N. sylvestris* (n = 12). From

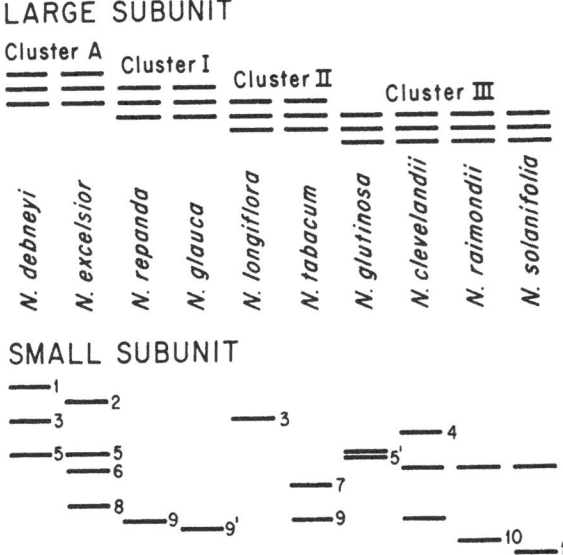

Fig. 4.2. Diagrammatic representation of the four different clusters of large subunit polypeptides and thirteen different kinds of small subunit polypeptides resolved by electrofocussing rubisco obtained from 63 species in the genus *Nicotiana*. Adapted from Chen *et al.* (1976a).

electrofocussing analysis of rubisco, Gray *et al.* (1974) determined that the latter species was the female parent and also provided the nuclear coding information for small subunit polypeptide #7. Genetic information for polypeptide #9 in *N. tabacum* was originally contained in *N. tomentosiformis* pollen.

By comparing tryptic and chymotryptic peptide maps of rubisco small subunits, Kawashima *et al.* (1974) showed that the *N. tabacum* rubisco small subunit was different in peptide composition from that in either parent. However, the *N. tabacum* small subunit composition could be reproduced by complementation whereby peptides missing in one parent would be replaced by peptides present in the other. Strøbaeck *et al.* (1976) showed that the *N. tabacum* rubisco small subunit to differ from that obtained from *N. sylvestris* or *N. tomentosiformis* in the amounts of tyrosine, isoleucine, aspartic acid, glycine, valine and histidine. The differences in amino acid composition were accounted for by a small subunit composed of equal quantities of an *N. sylvestris* kind of polypeptide together with an *N. tomentosiformis* kind of polypeptide. Sequencing by the same investigators showed that isoleucine occupies residue 7 in the *N. sylvestris* small subunit whereas tyrosine occupies residue 7 in the *N. tomentosiformis* small subunit. In contrast, either tyrosine or isoleucine could occupy residue 7 in the small subunit of *N. tabacum* rubisco. However, the seeming ambiguity at residue 7 was eliminated by later knowledge that the *N. tabacum* rubisco small subunit was composed of two, discrete polypeptides, one containing tyrosine and the other isoleucine at residue 7. Since substitution of tyrosine for isoleucine requires more than one base change in a coding triplet, it is clear that a second polypeptide of different isoelectric point could not have arisen in *N. tabacum* rubisco by a simple point mutation in the genetic code for an already existing polypeptide. Rather, the origin of *N. tabacum* rubisco is illustrative of the fact that higher plants have developed a mode of protein evolution where a large number of changes in amino acid composition can occur in conjunction with a single evolutionary event. It is furthermore possible for the plant breeder to create new rubiscos with predictable amino acid compositions by conventional interspecific hybridization or by fusion of protoplasts from different species of plants.

Another example of dramatic change in composition of the rubisco small subunit created by amphiploidy is illustrated by the *N. glauca* + *N. langsdorffii* parasexual hybrids produced by Smith *et al.* (1976) by fusion of protoplasts. *N. glauca* rubisco has a small subunit composed of a polypeptide #9′ whose isoelectric point is more acidic than polypeptides #3 and #5 which compose the *N. langsdorffii* rubisco small subunit. Each of the 16 parasexual hybrids contained a new species of rubisco with small subunits composed of polypeptides #3, #5, and #9′ (Chen *et al.*, 1977). Rubiscos containing the same three kinds of small subunit polypeptides were found in 46 F_2 progeny of self-pollinated parasexual hybrids and in eight self-pollinated F_3 progeny.

After a new species of rubisco has evolved by amphiploidy, what is remarkable is the faithfulness by which the new composition is preserved through generation after generation of plants. This can be shown by other examples in addition to the parasexual hybrids.

Constancy in composition of rubiscos containing more than one kind of small subunit polypeptides

Analysis of more than 100 different rubisco preparations from individual progeny of self-pollinated *N. tabacum* plants has revealed no difference in the isoelectric points of the two polypeptides. The two kinds of small subunit polypeptides stain with about equal intensity with bromophenol blue and no conspicuous change in this property has been noticed. Likewise, total amino acid analyses of the large and small subunits of different preparations of *N. tabacum* rubisco made by different laboratories are the same within limits of experimental error (Kawashima *et al.*, 1971; Strøbaeck *et al.*, 1976). It seems therefore that the amount of each of the two polypeptides initially combined during the phenomenon of amphidiploidy has remained constant in subsequent generations of tobacco plants.

An additional example of constancy in composition is that of tomato (*Lycopersicon esculentum*) whose rubisco contains three kinds of small subunit polypeptides (Gatenby and Cocking, 1978; Uchimiya *et al.*, 1979). Numerous accessions of wild cultivars of tomato had rubisco indistinguishable in polypeptide composition from rubisco in domesticated cultivars.

Genetic basis for constancy in rubisco composition

From what is now known about rubisco genetics, the simplest explanation to account for constancy in composition is that the coding information for the large subunit in a given species of plant has no opportunity to undergo interchange with dissimilar information because the code is contained in chloroplast DNA transmitted exclusively by the maternal parent. Constancy in composition of the small subunit can be attributed to the coding information for each kind of polypeptide being sequestered on a chromosome which does not pair during meiosis with a chromosome containing dissimilar information. Several examples support this contention.

In the case of two diploid *Nicotiana* species, Gerstel and Burns (1974) found *N. otophora* x *N. tomentosiformis* to be self-fertile. *N. otophora* (n = 12) rubisco has a single small subunit polypeptide (#7, Figure 4.2) whereas *N. tomentosiformis* (n = 12) rubisco has the single polypeptide #9. The *N. otophora* x *N. tomentosiformis* rubisco has a small subunit composed of polypeptides #7 and #9 in equal proportions, but each present to one-half the amount of that in either parent. The rubisco composition of 16 F_2 progeny of a selfed *N. otophora* x *N. tomentosiformis* F_1 hybrid distributed in the simple Mendelian ratio of 1 *N. otophora* type : 2 F_1 hybrid type : 1 *N. tomentosiformis* type (Chen and Wildman, 1980b). Evidently, the information for each kind of rubisco small subunit polypeptide was contained on heterologous chromosomes which segregated in typical diploid fashion during meiosis.

The presence and amounts of polypeptides #7 and #9 composing the small subunit polypeptide of *N. tabacum* rubisco is not affected by gene dose, the composition being the same for rubiscos extracted from haploid, diploid, triploid and tetraploid *N. tabacum* plants (Chen *et al.*, 1975b). A complete monosomic series exists where one chromosome is missing for each of the 24 different pairs of *N. tabacum* somatic chromosomes. While it is still not known which of the pairs of chromosomes (or perhaps all of them) contain the small subunit coding information, absence of one chromosome of a pair also does not affect the polypeptide composition of rubisco (Chen and Wildman, 1980b) which accords with expectation for polyploid behaviour.

As mentioned earlier, the genetic information for three kinds of rubisco small subunit polypeptides contained in polyploid *N. glauca* + *N. langsdorffii* parasexual hybrids did not segregate in F_2 and F_3 progeny. This condition affords another example of how rubisco composition in polyploids is preserved because heterologous chromosomes do not interchange genetic information during alternation of generations.

Evolution of the 13 different kinds of polypeptides composing the small subunits of *Nicotiana* rubiscos

Of the 29 different species of rubisco which are now extant among 65 species of *Nicotiana*, 60% have small subunits with more than one kind of small subunit polypeptide (Chen and Wildman, 1980a). Evolution of *Nicotiana* species also produced rubisco small subunit polypeptides with different isoelectric points of which 13 kinds are now existent. Analysis of rubisco in *N. suaveolens*, an amphiploid known previously for polymorphism in phenotype (Goodspeed, 1954), provides an indication of how small subunit polypeptides of differing isoelectric points evolved.

By random sampling of rubisco composition of individual plants in a population of *N. suaveolens*, three races were distinguished on the basis of different compositions of rubisco small subunits. Race 'A' and race 'B' had rubisco small subunits composed of three kinds of polypeptides. Polypeptides #6 and #8 (corresponding in relative isoelectric points to those shown in Figure 4.2) had the same isoelectric points and were in the same amounts in race A and race B rubisco. Race A contained polypeptide #2 while race B rubisco contained polypeptide #5. Race 'C' had a rubisco with all four kinds of polypeptides but #2 and #5 were reduced in amount by one-half compared with the amount of either one in race A or race B rubisco.

The race A and race B rubisco compositions bred true in F_2 progeny of self-pollinated plants but race C rubisco did not. Instead, race C yielded F_2 progeny containing race A, race B, and race C rubiscos in the ratio of 1 : 1 : 2 respectively. Apparently, genes for small subunit polyptptides #2 and #5 were allelic and segregated without dominance as a factor. However, in the same progeny, genetic information for polypeptides #6 and #8 did not segregate and behaved as if located on heterologous chromosomes in the same manner noted before for other *Nicotiana* polyploids.

Race C rubisco composition, consisting of four small subunit polypeptides, was recreated in reciprocal F_1 hybrids by crossing plants containing race A or race B rubiscos. Since race C already existed in a natural population of *N. suaveolens* which included race A and race B, the allelic character of genes coding for polypeptides #2 and #5 could signify that #2 arose by point mutation from #5, or *vice versa*, giving rise to a new race of rubisco which could be either A or B. Perpetuation of the new race of rubisco in F_2 progeny would have occurred by segregation at meiosis of the homologous chromosome pair bearing the mutated allele. The mutated *N. suaveolens* small subunit polypeptide provides direct support for a view previously presented (Uchimiya *et al.*, 1977) that the 13 polypeptides of different isoelectric points now extant among species of *Nicotiana* arose by point mutations.

Rubisco isozyme mixtures which develop as a consequence of evolution by amphiploidy

When Hirai (1977) subjected undenatured rubiscos from different species of plants to two-dimensional, agarose gel electrophoresis, the approximately one-half million dalton macromolecules migrated either as homogeneous proteins with respect to electrical charge, or as a heterogeneous group of macromolecules. Homogeneity was exhibited when the small subunit was composed of a single kind of polypeptide. Heterogeneity was displayed when the small subunit was composed of two kinds of polypeptides of different isoelectric points as detected by electrofocussing. The heterogeneity was interpreted as the consequence of random withdrawal of one or the other of the two kinds of polypeptides from a pool during the process of assembling eight small subunit polypeptides with eight large subunits to produce a finished rubisco macromolecule. Those rubiscos having only one or the other kind of polypeptide would appear at a frequency of 0.37% each. The remaining 99.3% of the rubisco molecules would contain small subunits composed of the two kinds of polypeptides in all possible combinations, the most frequent (27%) being small subunits composed of equal portions of the two kinds of polypeptides. Thus, rubisco seems to exist as a very subtle mixture of isozymes of different electrical charge when the small subunits are composed of more than one kind of polypeptide of different isoelectric points. But the unusual feature of rubisco isozymes is the constancy in composition of the mixtures compared to the changing composition in the mixtures of other kinds of plant isozymes. The *N. tabacum* rubisco mixture of nine isozymes has exhibited the same electrofocussing composition of large and small subunits irrespective of the developmental stage of the plant, the season of the year, the nutritional status of the plant, whether the protein was isolated from the white or the green portion of variegated leaves, or whether the rubisco had been synthesized in leaves never exposed to light. The isozyme composition has remained unaltered in more than 20 self-fertile cultivars which have been under constant manipulation by plant breeders for many generations and whose morphological phenotypes consequently exhibit great differences. The behaviour of rubisco isozymes in tobacco is therefore in sharp contrast to peroxidase in the same plant which separates into 12 distinguishable proteins of different electrophoretical mobilities. The proportion of the different electrophoretical species of peroxidase is variable in regard to the amount of each protein moiety at different stages of development of the tobacco plant (Sheen, 1970).

Reichenbächer *et al.* (1978) noted the unusual stability in mobility of *Hordeum vulgare* rubisco over a 35-day growth period. Electrofocussing has resolved only a single kind of rubisco small subunit polypeptide (Chen *et al.* 1976a) which indicates the barley protein to be homogeneous and not a mixture of isozymes. However, the stability in mobility is further assurance that comparable stability can be expected for *Triticinae* rubiscos. In contrast to *Hordeum* rubisco, I would expect that rubisco isozymes may have developed during tetraploid and hexaploid wheat evolution.

Rubisco behaviour in Dicotyledons as a model for further investigation of rubisco evolution in the *Triticinae*

Judging by what has happened to the composition of rubisco during evolution of species of *Nicotiana*, *Lycopersicon*, and other genera of Dicotyledons, it seems highly unlikely that the small subunit of *Triticinae* rubisco could have escaped modification in chemical composition during a course of evolution that led to the genetic complexity of hexaploid wheats. Reasons have been advanced for suspecting that modification in chemical composition of the rubisco large subunit is much more conservative than modification of the small subunit (Chen *et al.*, 1976b; Chen and Wildman, 1980a). *Triticinae* rubiscos have large subunits whose polypeptide clusters are of two types, an indication that rubisco has had an extensive period of time for evolutionary change. Why, then, has there been no evolution in composition affecting the isoelectric point of the small subunit? The answer may be that a change in composition accompanied by a change in isoelectric point cannot as yet be predicted on theoretical gounds. For example, Martin (1979) sequenced the 120 amino acids composing the rubisco small subunit obtained from spinach. Only a single ambiguity appeared at residue 91 where either tyrosine or proline could occupy this position. Yet this single difference resulted from the presence of two kinds of polypeptides which Chen *et al.* (1976b) showed could be resolved by electrofocussing. In contrast, the two kinds of polypeptides in the *N. tabacum* rubisco small subunit do not separate from each other to a much greater degree than the two spinach polypeptides even though the tobacco polypeptides differ in the amounts of seven amino acids. I would imagine therefore that similarity in small subunit isoelectric points exhibited by *Triticinae*rubiscos may not be a satisfactory index of homogeneity. On this account, I propose that a condition illustrated by the diagram in Figure 4.3 might pertain to rubisco evolution among wheat

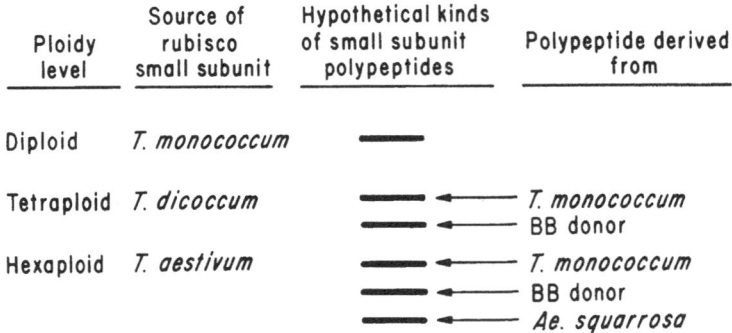

Ploidy level	Source of rubisco small subunit	Hypothetical kinds of small subunit polypeptides	Polypeptide derived from
Diploid	*T. monococcum*		
Tetraploid	*T. dicoccum*		*T. monococcum* BB donor
Hexaploid	*T. aestivum*		*T. monococcum* BB donor *Ae. squarrosa*

Fig. 4.3. Hypothetical scheme illustrating possible combination of different kinds of small subunit polypeptides during evolution of tetraploid and hexaploid wheats.

species. *T. monococcum* might possess a rubisco small subunit composed of a single polypeptide with no ambiguities in sequence. When *T. monococcum* hybridized with the BB genome donor (whatever species it may have been), I presume that at least one difference in amino acid sequence existed between the BB donor and *T. monococcum* rubisco small subunits with the high probability that tetraploid *T. dicoccum* rubisco has a small subunit composed of more than one kind of polypeptide. Extending the argument, hexaploid wheat is likely to contain rubisco small subunits comprised of three or more polypeptides each different in amino acid sequence. The hypothesis presented in Figure 4.3 can be tested by isolating native rubisco in pure form and subjecting the protein to the following kinds of analyses:

1. *Electrophoresis*

As shown by the diagram in Figure 4.4, isozyme mixtures of *Nicotiana* rubisco have been identified by their behaviour during two-dimensional, agarose gel electrophoresis. Homogeneous rubisco maintains a circular outline surrounding its position after electrophoresis whereas rubisco isozymes spread to cause an ellipsoidal outline to appear. With *Triticinae* rubisco, even very slight differences in mobilities might be enough to distort the circular outline and thereby suggest that rubisco isozymes could be present.

Additional experimentation would be predicated on dissociating rubisco before separating the large from the small subunits so that the latter could be subjected to further analysis.

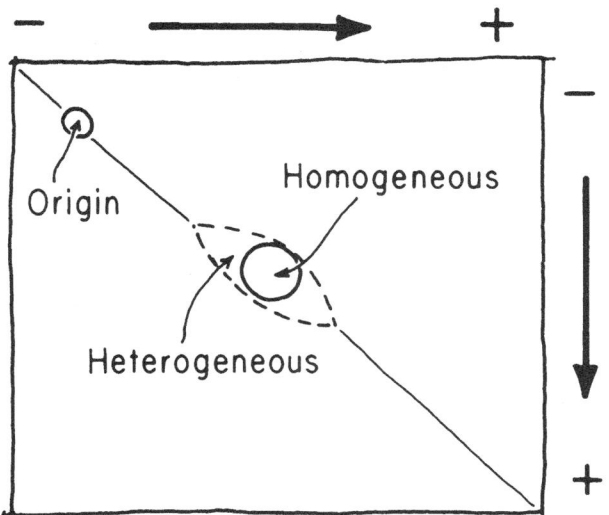

Fig. 4.4. Diagram illustrating expected result of subjecting undenatured rubisco to agarose gel electrophoresis in two dimensions. Homogeneous: rubisco small subunit consists of a single kind of polypeptide. Heterogeneous: small subunit composed of more than one kind of polypeptide of different charge so that rubisco becomes a mixture of isozymes with slight differences in mobilities. Adapted after Hirai (1977).

2. *UV absorption spectra*

Gray *et al.* (1978) separated the two kinds of polypeptides composing the *N. tabacum* rubisco small subunit, dissolved them in 0.1 N Na OH and examined their ultraviolet spectra in the region of tyrosine and tryptophan absorption. Polypeptide #7 (Figure 4.2) had a tyrosine/tryptophan ratio of 1.5 : 1 whereas the ratio for polypeptide #9 was 2.5 : 1. Without fractionation of the polypeptides, the small subunit of *N. tabacum* rubisco had an intermediate ratio of 2.0 : 1. If changes in tyrosine or tryptophan have occurred during evolution of rubisco in *Triticinae*, they might be detected by UV absorption and provide clues as to where heterogeneity in kinds of polypeptides might be encountered.

3. *Amino acid analysis*

Comparison of the total amino acid composition of the rubisco contained in diploids *vs.* tetraploids, tetraploids *vs.* hexaploids, diploids *vs.* hexaploids, etc., might reveal differences and point to where heterogeneity in polypeptide composition could be expected.

4. Sequence analysis

With automatic sequencing, the order of the first 25 amino acids from the N-terminal end of the small subunit polypeptide can be obtained with accuracy. It is possible that ambiguities might appear in this region as in the case of the two *N. tabacum* rubisco small subunit polypeptides.

5. Cyanogen bromide cleavage

In most species of rubisco, methionine appears only as the N-terminal amino acid of the small subunit. However, additional methionine may be present as in the sequence of the small subunit of spinach rubisco. Cyanogen bromide cleaves peptide bonds where methionine appears and this reagent might produce two or more peptide fragments in *Triticinae* rubisco small subunits whereby diploid, tetraploid, or hexaploid rubiscos might be distinguished on the basis of molecular weight, amino acid composition or electrofocussing differences and thus provide evidence for the presence of more than one kind of rubisco small subunit polypeptide.

6. Tryptic and chymotryptic peptide fingerprinting

Kawashima *et al.* (1974) subjected the small subunits of *N. tomentosiformis*, *N. sylvestris*, and *N. tabacum* rubisco small subunits to chymotryptic peptide fingerprinting. They found the *N. tabacum* fingerprint to contain one peptide not present in *N. tomentosiformis*, and another not present in *N. sylvestris* small subunits. If similar type of evidence were obtained for *Triticinae* rubisco small subunits, this would be strong presumptive evidence that more than one kind of polypeptide exists.

Identification of small subunits composed of more than one kind of polypeptide among *Triticinae* rubiscos would open up new avenues for exploration of wheat evolution. An immediate progenitor of the BB genome should contain a rubisco whose small subunit corresponds in amino acid sequence with one of the kinds of polypeptides expected to be found in *T. dicoccum*. Barring discovery of an existing accession with a rubisco small subunit composition which corresponds with expectation for direct descent, degrees of relatedness could be assigned on the basis of how closely the amino acid sequence corresponds to one of the kinds of polypeptides constituting the small subunit of tetraploid and hexaploid wheats.

References

Chen, K. and Wildman, S.G. (1980a) Inheritance behaviour of information coding for small subunit polypeptides of Fraction 1 protein. *Biochem. Genet.*, (in press).

Chen, K. and Wildman, S.G. (1980b) Differentiation of Fraction 1 protein in relation to age and distribution of Angiosperm groups. *Plant Syst. Evol.* (in press).

Chen, K., Gray, J.C., and Wildman, S.G. (1975a) Fraction 1 protein and the origin of polyploid wheats. *Science* **190**, 1304–1306.

Chen, K., Kung, S.D., Gray, J.C. and Wildman, S.G. (1975b) Polypeptide composition of Fraction 1 protein from *Nicotiana glauca* and from cultivars of *Nicotiana tabacum*, including a male sterile line. *Biochem. Genet.* **13**, 771–778.

Chen, K., Kung, S.D., Gray, J.C. and Wildman, S.G. (1976a) Subunit polypeptide composition of Fraction 1 protein from various plant species. *Plant Sci. Lett.* **7**, 429–434.

Chen, K., Johal, S. and Wildman, S.G. (1976b) Role of chloroplast and nuclear genes during evolution of Fraction 1 protein. In 'Genetics and Biogenesis of Chloroplasts and Mitochondria'. (Eds. T. Bucher, W. Neupert, W. Sebald and S. Werner.) North-Holland, Amsterdam, New York. pp. 99–102.

Chen, K., Wildman, S.G. and Smith, H.H. (1977) Chloroplast DNA distribution in parasexual hybrids as shown by polypeptide composition of Fraction 1 protein. *Proc. Natl. Acad. Sci. USA* **74**, 5109–5112.

Coen, D.M., Bedbrook, J.R., Bogorad, L. and Rich, A. (1977) Maize chloroplast DNA fragment encoding the large subunit of ribulose bisphosphate carboxylase. *Proc. Natl. Acad. Sci. USA* **74**, 5487–5491.

Gatenby, A. and Cocking, E.C. (1978) The polypeptide composition of the subunits of Fraction 1 proteins in the genus *Lycopersicon. Plant Sci. Lett.* **13**, 171–176.

Gerstel, D.U. and Burns, J.A. (1974) Meiosis and staminal sterility in hybrids between *Nicotiana tomentosiformis* and *N. otophora. Tob. Sci.* **18**, 157–159.

Goodspeed, T.H. (1954) 'The genus *Nicotiana.*' Chronica Botanica Press, Waltham, Mass.

Gray, J.C., Kung, S.D., Wildman, S.G. and Sheen, S.J. (1974) Origin of *Nicotiana tabacum* L. detected by polypeptide composition of Fraction 1 protein. *Nature* **252**, 226–227.

Gray, J.C., Kung, S.D. and Wildman, S.G. (1978) Polypeptide chains of the large and small subunits of Fraction 1 protein from tobacco. *Arch. Biochem. Biophys.* **185**, 272–281.

Hirai, A. (1977) Random assembly of different kinds of small subunit polypeptides during formation of Fraction 1 protein macromolecules. *Proc. Natl. Acad. Sci. USA* **74**, 3443–3445.

Kawashima, N., Kwok, S.-Y. and Wildman, S.G. (1971) Studies on Fraction 1 protein. III. Comparison of the primary structure of the large and small subunits obtained from five species of *Nicotiana. Biochim. Biophys. Acta* **236**, 578–586.

Kawashima, N., Tanabe, Y. and Iwai, S. (1974) Similarities and differences in the primary structure of Fraction 1 proteins in the genus *Nicotiana. Biochim. Biophys. Acta* **371**, 417–431.

Martin, P.G. (1979) Amino acid sequence of the small subunit of ribulose-1,5-bisphosphate carboxylase from spinach. *Aust. J. Plant Physiol.* **6**, 401–408.

Reichenbächer, D., Richter, J. and Spaar, D. (1977) Unterschiede in der electrophoretischen Beweglichkeit von Fraktion-I-Protein bei Gramineen und ihre mögliche Nutzung in Genetik und Züchtungsforschung. *Biochem. Physiol. Pflanz.* **171**, 299–306.

Reichenbacher, D., Borner, Th. and Richter, J. (1978) Untersuchungen am Fraktion-I-Protein der Gerste mit Hilfe quantitativer Immunelektrophoresen. *Biochem. Physiol. Pflanz.* **172**, 53–60.

Sakano, K., Kung, S.D. and Wildman, S.G. (1974) Identification of several chloroplast DNA genes which code for the large subunit of *Nicotiana* Fraction 1 proteins. *Mol. Gen. Genet.* **130**, 91–97.

Sheen, S.J. (1970) Peroxidases in the genus *Nicotiana. Theor. Appl. Genet.* **40**, 18–25.

Smith, H.H., Kao, K.N. and Combatti, N.C. (1976) Confirmation and extension of interspecific hybridization by somatic protoplast fusion in *Nicotiana. J. Hered.* **67**, 123–128.

Strøbaeck, S., Gibbons, G.C., Haslett, B., Boulter, D. and Wildman, S.G. (1976) On the nature of the polymorphism of the small subunit of ribulose 1,5-diphosphate carboxylase in the amphidiploid *Nicotiana tabacum. Carlsberg Res. Commun.* **41**, 335–343.

Uchimiya, H., Chen, K. and Wildman, S.G. (1977) Polypeptide composition of Fraction 1 protein as an aid in the study of plant evolution. *Stadler Genet. Symp.* **9**, 83–100.

Uchimiya, H., Chen, K. and Wildman, S.G. (1979) Evolution of Fraction 1 protein in the genus *Lycopersicon. Biochem. Genet.* **17**, 333–341.

Vedel, F., Quetier, F., Dosba, F. and Doussinault, G. (1978) Study of wheat phylogeny by EcoRI analysis of chloroplastic and mitochondrial DNAs. *Plant Sci. Lett.* **13**, 97–102.

5 TRANSFER OF ALIEN GENETIC MATERIAL TO WHEAT

E.R. Sears

As pointed out by Frankel (1970) and others, the pool of potentially useful genes available to wheat breeders has shrunk alarmingly in recent years, primarily because of the replacement of the highly variable land races in many parts of the world by higher-yielding, pure-line varieties. It appears that much of the genetic variability of the cultivated wheats has already been lost and cannot be recovered.

The picture becomes much brighter, however, when we realize that the relatives of wheat are accessible sources of genes for use in wheat improvement. The cultivated wheats are blessed with a large assortment of relatives (Table 5.1), diverse in phenotype and adaptation. They bring forth the pleasant prospect that wheats may be produced which are adapted to colder, hotter, drier, saltier, or less fertile environments; which have new genes for resistance; or which even have greater yielding capacity than existing cultivars when grown in present wheat-producing regions. Obviously, the more distant from wheat the relative is, the more likely it is to have genes that are not present in any of the wheats themselves. Some of these genes may be of great value to wheat growers.

Nearly all of the closer relatives can be crossed with wheat, especially if advantage is taken of such procedures as searching for crossable biotypes, stimulating seed development with growth hormones, and culturing embryos on artificial media. Even barley, which belongs to a different sub-tribe than wheat, has now been crossed with wheat (Kruse, 1973; Islam *et al.*, 1975; Fedak, 1977). Should any blocks to crossing be found, the desired hybrids can presumably be obtained by using the techniques of somatic-cell fusion, once these have been perfected.

Until the past 20 years, an almost insurmountable block to transfer of genes to wheat was the inability of wheat chromosomes to pair with those of any of its relatives except those few that have one or more genomes (sets of seven pairs of chromosomes) homologous with wheat genomes. Crosses were able to be made and hybrids grown, but failure of chromosome pairing all but precluded the transfer of alien genetic material to wheat chromosomes. To be sure, it was known that exchanges could be induced with ionizing radiation, and several genes for disease resistance were transferred to wheat in this way (reviewed by

Table 5.1

Groups of wild relatives of wheat, listed in decreasing order of their presumed closeness of relationship to common wheat (ABD)

Type of gene pool	Species and genomic formulae
1. Species with homologues of wheat genomes	
(a) The tetraploid progenitor	*T. turgidum* var. *dicoccoides* (AB)
(b) The diploid donors of the A and the D genomes	*T. monococcum* var. *boeoticum* or var. *urartu* (A) *T. tauschii* (D)
(c) Polyploids with one homologous genome	
(i) The A genome	*T. timopheevii* var. *araraticum* (AG)
(ii) The D genome	*T. crassum* (DM^{cr}, DD_2M^{cr}) *T. ventricosum* $(DM^v)^2$ *T. cylindricum* (CD) *T. juvenale* $(DM^{cr}U)$ *T. syriacum* $(DM^{cr}S)$
2. Species with homoeologous genomes	
(a) Closely related species	*T. searsii* (S^s) *T. longissimum* (S^1) *T. sharonensis* (S^1) *T. bicorne* (S^b) *T. speltoides* (S) *T. variabile* (US^v) *T. kotschyi* (US^v)
(b) Less closely related species	*T. tripsacoides* (Mt) *T. dichasians* (C) *T. comosum* (M) *T. umbellulatum* (U) *T. uniaristatum* (M^u) Other U-containing polyploids Several *Agropyron* species
(c) Distantly related species	Species of *Secale, Haynaldia*; numerous species of *Agropyron* and of other genera of the *Triticinae* and *Hordeinae*

Knott, 1971); but this is at best a very laborious process, with a low probability of yielding an acceptable transfer.

The failure of wheat chromosomes to pair with those of related species severely restricted the exploitation of the relatives. The chromosome number of a hybrid of wheat with a relative could be doubled, resulting in a fertile amphiploid, but these, with one exception, have never been commercially successful. The exception, *Triticale*, has up to seven rye pairs added to the 14 chromosome pairs of tetraploid wheat; but the rye chromosomes are not from a typical relative, which would be a wild species, but are from a cultivated species.

It is also possible to add almost any pair of alien chromosomes to wheat, producing an alien-addition line. This rarely, if ever, leaves the genotype well balanced. Also, loss of the alien chromosome is favoured by pollen selection, with the result that the line tends to return to the euploid condition.

A better stratagem for introducing alien variation is to substitute the alien chromosome pair for a homoeologous (related) pair of wheat chromosomes. This is easily accomplished by crossing the addition line onto the proper wheat monosomic and recovering an F_2 plant deficient for the wheat pair and disomic for the alien pair. This plant and the descendent alien-substitution line may be essentially normal in phenotype and of stable constitution, like several East European wheat cultivars that have chromosome 1R of rye substituted for wheat chromosome 1B (Mettin *et al.*, 1973; Zeller, 1973). Nearly always, however, the complete alien chromosome carries with it one or more undesirable genes, or else it does not compensate satisfactorily for the missing wheat chromosome.

A further possibility has been identified in recent years (Sears, 1972a): to substitute one arm of an alien chromosome for a corresponding wheat arm. This can be achieved by taking advantage of the fact that univalents in wheat frequently misdivide at meiosis I. If both the alien chromosome and its wheat homoeologue are monosomic, both will be univalent at meiosis. When both happen to misdivide in the same sporocyte and give rise to telocentric chromosomes, these will occasionally fuse at the centromeres to produce a bibrachial chromosome with one arm from the alien chromosome and one from the homoeologue. The frequency with which the correct combination arises is so low, however, that the amount of effort required is little, if any, less than would be needed to induce homoeologous pairing and replace only part of the wheat arm with the corresponding alien segment. Therefore whole-arm substitution is not likely to become a popular method for introducing alien variation.

Of great potential value to wheat breeding was the discovery by Okamoto (1957) and Riley and Chapman (1958; also Riley *et al.*, 1958) that the failure of related chromosomes to pair is due to a particular wheat chromosome, 5B. In the absence of 5B, not only homologues but also homoeologues can pair; but if both modes are possible, homologous pairing is favoured over homoeologous. High levels of homoeologous pairing usually occur in hybrids that lack 5B (Figure 5.1).

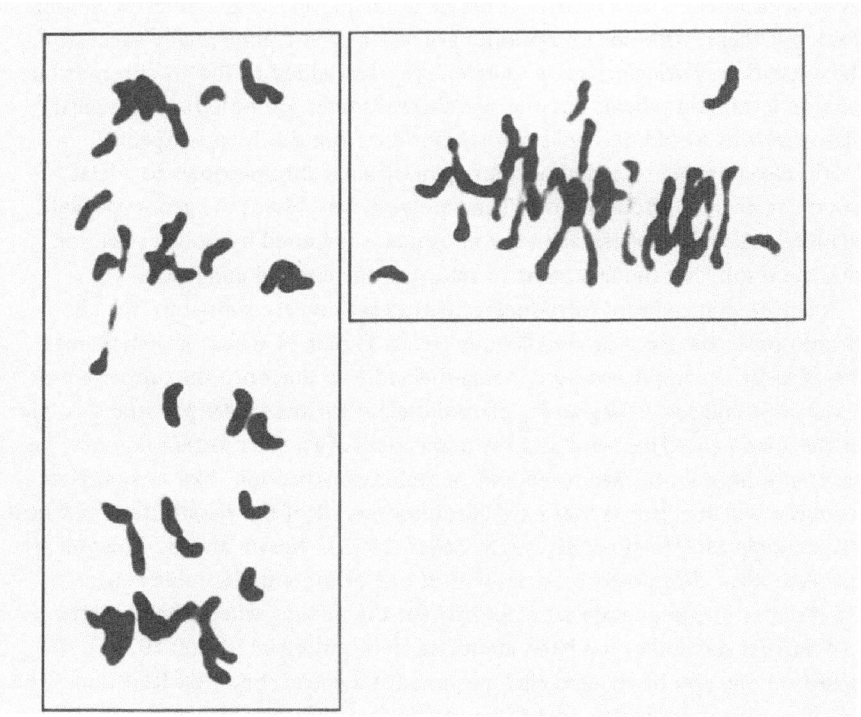

Fig. 5.1. Effect of the suppressor of homoeologous pairing, *Ph1*, on meiosis in the hybrid, hexaploid wheat (AABBDD) x *Triticum kotschyi* (UUS^{V}S^V). With *Ph1* present (left), 33 chromosomes are unpaired and only 2 paired, whereas with *Ph1* absent (right), only 6 chromosomes are unpaired and 29 paired.

To induce alien chromosomes to pair with their wheat homoeologues, it is only necessary to delete chromosome 5B. The simplest way to do this is to pollinate monosomic 5B by the desired alien species. About 75% of the offspring will be deficient for chromosome 5B. Riley (1966b) used this method to transfer genetic material from *Triticum bicorne (Aegilops bicornis)* to wheat, and Joshi and Singh (1979) succeeded in introducing rye genes into wheat in the same way. Instead of mono-5B, nullisomic-5B tetrasomic-5D could have been used, or, better, a mutant line *ph1b ph1b* (Sears, 1977) evidently deficient for the pairing suppressor *Ph1* (for pairing homoeologous).

A problem encountered with 5B- and *Ph1*-deficient hybrids is their extremely low fertility. The probability of recovering a desirable transfer of a particular alien gene to wheat is very low if only a few seeds are obtained on the F_1, for most transfers involve a relatively long alien segment, likely to carry deleterious genes as well as the one desired. The almost complete sterility of the F_1 may be attributed to the high level of pairing, which presumably reduces the frequency of failure of the first meiotic division and thereby cuts down on the formation of restitution nuclei.

The fertility should be better if a less effective pairing mutant were used — either the partial mutant *ph1a* obtained by Wall *et al.* (1971) or Sears' (1977) *ph2* mutant on 3DS. However, the increased fertility would be offset by a lower frequency of recombination between homoeologues; thus there might be no net gain unless the alien genome happened to be especially closely related to one of the wheat genomes. In that case, as Feldman and Sears (1981) have suggested, the intermediate-pairing mutant might permit good pairing between the closely related genomes while largely suppressing pairing of the other homoeologues.

Rather than deleting *Ph1*, it is possible to suppress its inhibitory activity by adding the chromosomes of certain strains of *T. speltoides* (*Ae. speltoides*) or *T. tripsacoides* (*Ae. mutica*). In simple hybrids of wheat with these strains, homoeologous pairing occurs (Riley *et al.* 1958; Riley, 1966a) and can result in the transfer of *speltoides* or *tripsacoides* segments to wheat chromosomes. Further, when a high-pairing wheat-*speltoides* (or -*tripsacoides*) amphiploid is crossed with another alien species, homoeologous pairing occurs in the F_1. However, the *speltoides* chromosomes can pair and recombine with the wheat chromosomes, and also with those of the other alien species, thereby adding to the difficulty of sorting out the desired transfers. Nevertheless, Riley *et al.* (1968) succeeded in recovering in this way a part-wheat, part-*T. comosum* chromosome that provides resistance to the stripe-rust fungus.

For the transfer of a particular gene from an alien species to wheat, a degree of simplification, and at the same time greater precision, can be achieved by first isolating the critical alien chromosome by producing an addition line. This will be a line with the complete complement of wheat chromosomes plus one alien pair. Such lines are relatively easy to obtain (Figure 5.2) and in fact already exist for most or all of the chromosomes of a number of relatives, including rye, several diploid *Triticum* species, *Haynaldia villosa*, *Agropyron elongatum*, and even barley. The alien-addition line can be converted to homoeologous pairing by introducing the *ph1b* mutation, or by making the line nullisomic-5B and tetrasomic, or trisomic, 5D (Sears, 1972a). If the alien chromosome is monosomic instead of disomic, it will pair more freely with its homoeologues; and if one of the wheat homoeologues is also monosomic, a still higher level of homoeologous pairing can be achieved. This has an advantage that may be substantial; namely, most or all of the resulting transfers will involve a particular wheat chromosome instead of being distributed among all three homoeologues (Sears, 1972a).

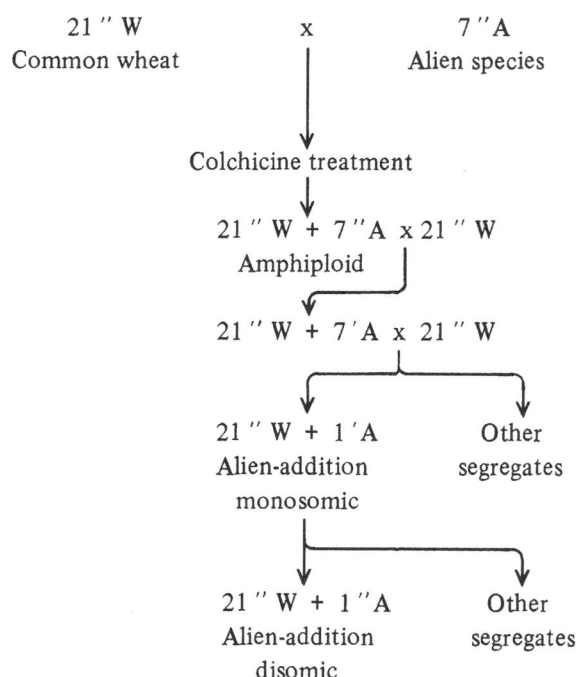

Fig. 5.2. Production of an alien-addition line. Alien species with 14 "may also be used. The colchicine treatment may be omitted, since the non-doubled hybrid pollinated by normal plants will usually give rise to a reasonable frequency of offspring with 21 "W + 7 'A, as the result of restitution-nucleus formation.

In order to make both an alien chromosome and its homoeologue monosomic, it is only necessary to obtain an alien substitution line (Figure 5.3) and cross it with a euploid. If the euploid is *ph1b ph1b* (or nulli-5B, tetra-5D) and the substitution line has been made monosomic-5B (Figure 5.4), the F_1 will be monosomic for both desired chromosomes and at the same time have homoeologous pairing (Figure 5.5).

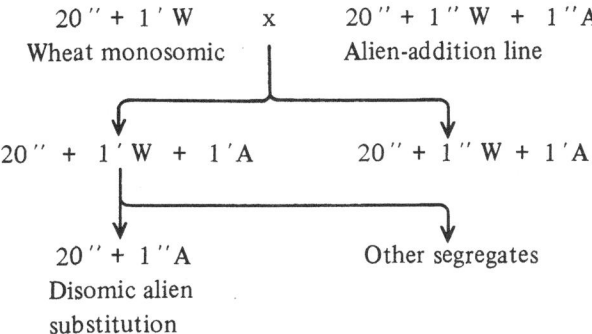

Fig. 5.3. Production of an alien-substitution line. From the plant with 20 ″+ 1 ′W + 1 ′A, a fair percentage of the functioning male gametes are expected to carry the alien chromosome instead of its wheat homoeologue.

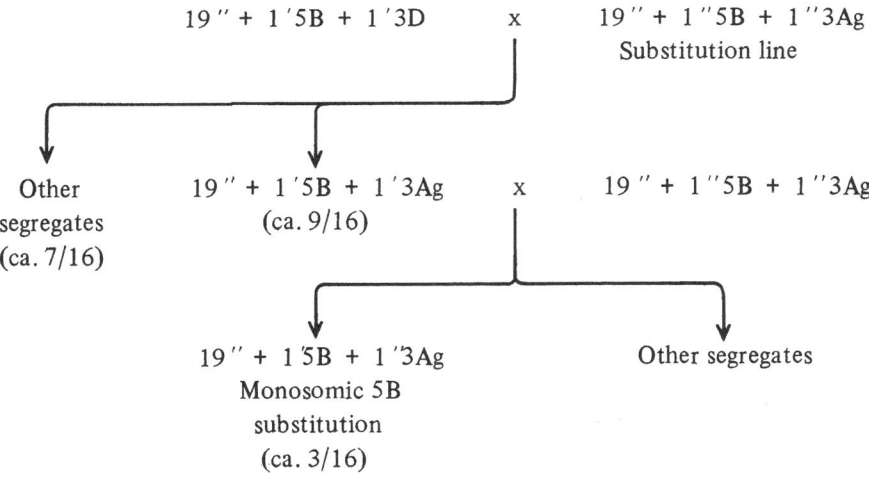

Fig. 5.4. One procedure for making a substitution line monosomic for chromosome 5B, as illustrated by the 3Ag(3D) substitution line.

Another step that may profitably be inserted into the procedure is to obtain a telocentric derivative (telosome) for the alien arm carrying the gene concerned and use this instead of the complete alien chromosome. The complete chromosome will have been monosomic at a previous stage and, since alien univalents tend to misdivide at a relatively high rate, a telosome for the proper arm may already have been recovered without a special search. One advantage of using the telosome is that it can easily be identified whenever it pairs with its homoeologue, thereby permitting a precise determination of the pairing frequency and the prediction of the resultant frequency with which transfer chromosomes (recombinant chromosomes carrying the specified alien gene) will be produced (about one-fourth of the pairing frequency). A greater advantage of using the telosome is that this facilitates the identification and characterization of the transfer chromosomes. A possible third advantage is that recombination tends to be shifted distally in the telosome as compared with the complete chromosome. Whether or not this shift will be advantageous depends on the location of the alien gene concerned and the distribution pattern of recombination between the alien chromosome and its wheat homoeologue.

With the alien chromosome monotelosomic and its homoeologue monosomic, several types of gamete are produced (Figure 5.5) depending on whether recombination occurs proximal or distal to the gene concerned. Following a cross to euploid, the offspring carrying the alien gene will be of three types: (1) those with the unchanged alien telosome; (2) those with a telosome that has its terminal portion replaced by a wheat segment (as the result of a distal crossover); and (3) those with a wheat chromosome which, as the result of a proximal crossover, has had the distal portion of one of its arms replaced by an alien segment. These are easily distinguished cytologically, for in type 1 the telosome remains unpaired (*Ph1* being present); in type 2 it pairs, with a frequency depending on the length of the terminal wheat segment; and in type 3 the alien gene is present but there is no telosome.

The only experiment thus far completed using an alien telosome (Sears, unpublished) involved a chromosome (called 3Ag) from *Agropyron elongatum* carrying a gene *Lr24* for resistance to the leaf-rust fungus. A previous experiment, in which complete 3Ag and 3D were monosomic, 5B nullisomic, and 5D trisomic (Table 5.2), had given rise to 20 transfers among 299 offspring (Sears, 1972a, 1978). Two of the 20 involved 3B instead of 3D, and one involved both 3B and 3D. The latter was the only one to have resulted from 3Ag-3D exchange distal to *Lr24*. Its terminal 3DL segment was very short, only long enough to support about 3% pairing with 3D. The main reason for undertaking the experiment involving telo-3AgL was the possibility that crossing-over might be shifted distally, as is known to happen with homologously pairing telosomes (Endrizzi and Kohel, 1966; Sears, 1972b).

In the telo-3AgL experiment (Table 5.1) of 328 offspring, eight had transfer chromosomes, one of which was evidently 3B or 3A rather than 3D. Three were 3DL exchanges proximal to *Lr24* (capable of 2%, 11% and 48% pairing, respectively, with telo-3DL). The remaining four were the result of distal exchanges between telo-3AgL and 3DL. Three of the four had very short 3DL segments, none of which was able to support more than about 5% pairing; but the fourth

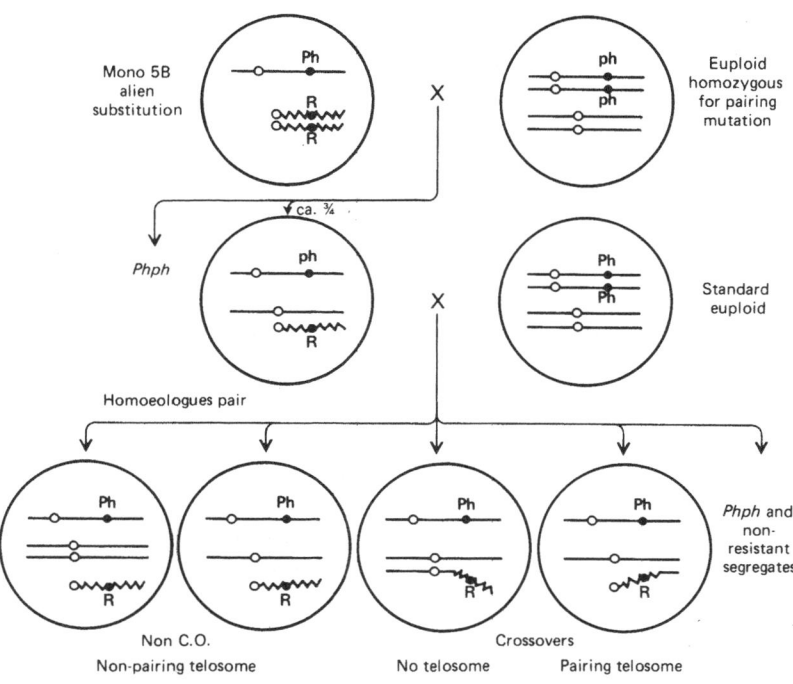

Fig. 5.5. Induction of pairing and crossing-over between an alien telosome and its wheat homoeologue. Each plant has 19 pairs of chromosomes in addition to those shown.

Table 5.2

Comparison of transfers induced from telosome 3AgL with those from the complete 3Ag chromosome

	Type of alien chromosome	
	Telosomic	Complete
Number of plants	328	299
Number with *Lr24*	91	77
Number of proximal transfers	3	17
Number of distal transfers	4	1*
Number not 3D	1	2

*Also had transfer with 3BS

had 28% pairing (in 50 cells). Thus the exchanges involving 3D consisted of 4 distal to *Lr24* and 3 proximal; whereas in the complete-3Ag experiment the corresponding numbers were 1 and 17. The fact that only eight transfers were obtained from telo-3AgL contrasts with the 18 of the previous, slightly smaller experiment and suggests that the increase in distal crossovers does not entirely compensate for the decrease in proximal exchange. This appears to be in contrast to Sallee and Kimber's (1979) finding that the frequency of homologous pairing of wheat telosomes is not significantly reduced, presumably because additional distal chiasmata compensate for the fewer proximal ones. However, the 18 transfer chromosomes include several that may well have an exchange in 3DS rather than 3DL.

In the foregoing experiment, telo-3DL and complete 3Ag might have been used instead of telo-3AgL and complete 3D, but not to as great advantage. Use of 3DL would have permitted scoring of the amount of its pairing with the alien arm; but identification of transfer chromosomes would have been more difficult than when 3AgL was used, mainly because the possible transfer chromosomes could not have been tested immediately for pairing with 3Ag, since combining them with 3Ag would have concealed the presence of *Lr24*.

The desirability of obtaining a transfer chromosome with a 3AgL segment shorter than any of those obtained in either of the two completed experiments is based on the reasoning that the shorter the segment of alien chromatin, the less the danger that one or more deleterious genes will accompany the desirable gene. This is not to deny that other advantageous genes may be present on a particular long, alien segment; in fact, the 3D/3Ag transfers thus far obtained carry a very desirable *Agropyron* stem-rust gene, *Sr24* (McIntosh, personal communication). But, barring prior knowledge of the existence of such additional genes, the cytogeneticist's goal must be to deliver to the breeder a transfer chromosome that includes the shortest possible alien segment.

The fact that one distal-transfer chromosome had 28% pairing with telo-3AgL clearly established that *Lr24* is located a substantial distance from the end of the arm. There was therefore virtually no chance of obtaining a transfer chromosome with a short alien segment in one episode of induced homoeologous pairing. Obtaining a short, interstitial segment requires two crossovers near together, and these are rare even between homologues. They are likely to be even less frequent between homoeologues.

The same result as from such a double crossover can be obtained, however, by combining two transfer chromosomes having exchanges near the gene on the proximal and distal sides, respectively (Figure 5.6). These two chromosomes are homologous only in the segment of alien chromatin they have in common, a region that includes the gene concerned. Every crossover that occurs will give rise to a chromosome that is entirely of wheat except for an interstitial segment consisting of the region possessed in common by the two parental chromosomes. Thus the closer each of the two original exchanges is to the desired alien gene, the shorter the interstitial segment will be in the derived recombinant chromosome.

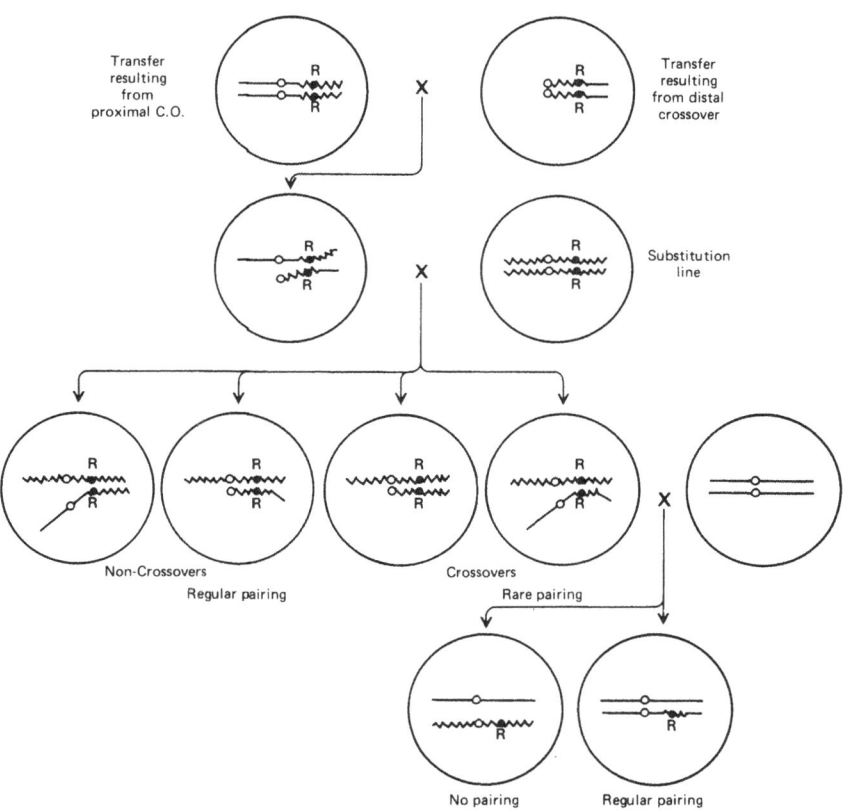

Fig. 5.6. Derivation and identification of a chromosome with a shortened interstitial alien segment from two transfers of different type.

Another possibility for obtaining a transfer chromosome with a short, intercalary alien segment carrying the desired gene is particularly applicable if a reasonably large experiment has yielded only one type of transfer; i.e., involving exchanges either all proximal or all distal to the gene concerned. In such case, the transfer chromosome with its exchange closest to the gene may be combined with the corresponding all-wheat chromosome in a plant whose genotype permits homoeologous pairing (Figure 5.7). Homoeologous recombination will then presumably occur between the corresponding alien and wheat segments, replacing part of the alien piece with wheat chromatin. Unfortunately, there will be difficulty in identifying the desired recombinants if the parental transfer chromosome already pairs in high frequency with the critical arm of the wheat chromosome.

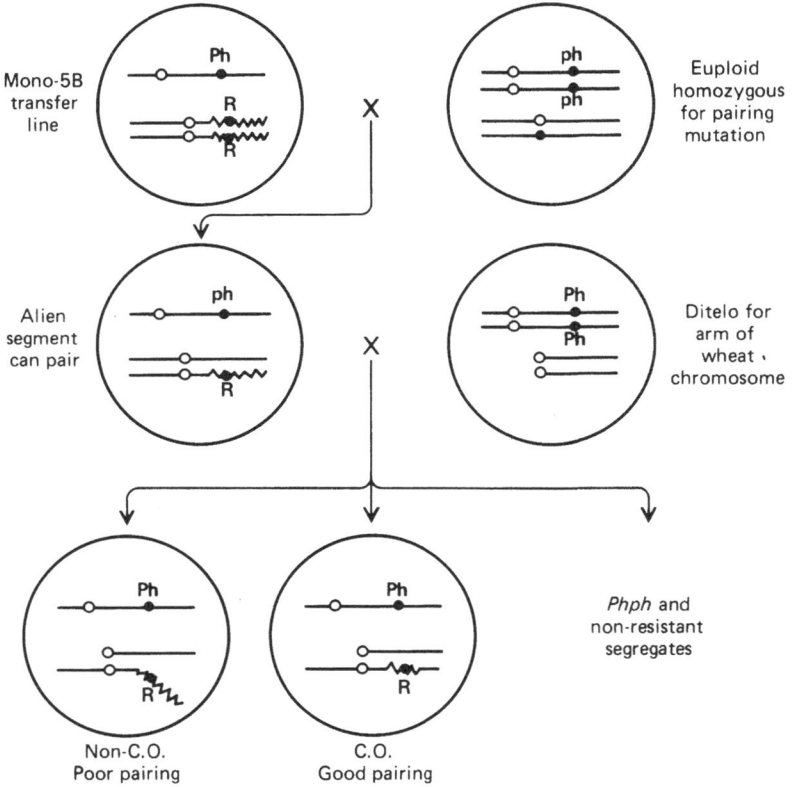

Fig. 5.7. Shortening the alien segment of a transfer chromosome by inducing it to pair homoeologously with the corresponding wheat segment.

If the experiment involving telo-3AgL had been the first attempt to transfer *Lr24* to wheat chromosome 3D, and if no transfer due to proximal recombination had been recovered, this would have signified that the gene was located in the proximal portion of the arm. An experiment using the complete 3Ag would then have been in order, because this would have tended to move the exchanges proximally. If *Lr24* had been tightly linked to the centromere, an exchange on the other side of the centromere in the short arm would still have been useful. In combination with a suitable exchange distal to *Lr24* in the long arm, it would have given rise to a 3D chromosome with an intercalary *Agropyron* segment carrying *Lr24*. The segment would have included the centromere, but this would presumably not have prevented the chromosome from behaving like any completely wheat chromosome, nor from pairing normally with 3D.

Of the 17 transfers resulting from the experiment involving the complete chromosome 3Ag, the two with the shortest *Agropyron* segment were transfers 3 and 14. Both had a relatively high frequency of pairing with telo-3DL (64% and 69%, respectively), and they are apparently the only transfers that have the *R1* (red seed) locus (which lies proximal to *Lr24*) from 3D instead of 3Ag (McIntosh, personal communication). When each of these two transfers was combined with the highest-pairing distal-transfer chromosome from the telo-3AgL experiment (the chromosome that paired with 3DL in 28% of cells), the segment that this chromosome had in common with transfers 3 and 14 proved to be substantial: enough to result in pairing in 89% of the cells.

The location of *Lr24* in this rather long region cannot be deduced from the available data. Obviously additional transfers must be induced from telo-3AgL. The experiment already performed with telo-3AgL resulted in the recovery of transfers with exchange points that were apparently fairly well distributed along the length of the arm. If a similar distribution occurs in a larger experiment, it is reasonable to expect some transfers with exchange points closer to *Lr24*, unless the segment in which *Lr24* is located is for some reason unable to support chiasmata with the corresponding portion of 3DL. If the desired transfer or transfers are obtained, they will not only locate *Lr24* more precisely but will permit the synthesis of a transfer chromosome with only a short interstitial *Agropyron* segment. In any case, the larger experiment should give a better idea of how nearly random the induced recombination is between 3AgL and 3DL.

Conclusion

Genes can be transferred to the chromosomes of wheat from its relatives by genetic induction of homoeologous pairing in hybrids, or derivatives from hybrids, that have one or more alien chromosomes. This can be done by removing chromosome 5B, on which *Ph1*, the major pairing suppressor, is located; by neutralizing *Ph1* with the genome of *Triticum speltoides* or *T. tripsacoides*; or by using the mutation *ph1b*, which is evidently a deficiency for the locus. A greater degree of precision can be attained by having present a single alien monosome and a monosome for one of its homoeologues. Having the alien chromosome telosomic facilitates identification and characterization of the transfer chromosomes obtained. By allowing recombination between two transfer chromosomes, one with its exchange point proximal to the alien gene concerned and the other distal, a wheat chromosome with an intercalated alien segment can be produced.

Acknowledgement

The research on which this paper is based was carried out while the author was employed by the U.S. Department of Agriculture.

References

Endrizzi, J. and Kohel, R.J. (1966) Use of telosomes in mapping three chromosomes of cotton. *Genetics* **54**, 535–550.

Fedak, G. (1977) Increased homoeologous pairing in *Hordeum vulgare* x *Triticum aestivum* hybrids. *Nature* **266**, 529–530.

Feldman, M. and Sears, E.R. (1981) Utilization of wild gene resources in the improvement of wheat. *Sci. Am.* (in press).

Frankel, O.H. (1970) Genetic conservation in perspective. In 'Genetic Resources in Plants – Their Exploration and Conservation'. (Eds. O.H. Frankel and E. Bennett.) Blackwell Scientific Publications. pp. 469–489.

Islam, A.K.M.R., Shepherd, K.W. and Sparrow, D.H.B. (1975) Addition of individual barley chromosomes to wheat. Proc. 3rd Int. Barley Genetics Symp. Garching, W. Germany. pp. 260–270.

Joshi, B.D. and Singh, D. (1979) Introduction of alien variation into bread wheat. Proc. 5th Int. Wheat Genetics Symp., New Delhi, 1978. pp. 342–348.

Knott, D.R. (1971) The transfer of genes for disease resistance from alien species to wheat by induced translocations. In 'Mutation Breeding for Disease Resistance'. International Atomic Energy Agency, Vienna. pp. 67–77.

Kruse, A. (1973) *Hordeum* x *Triticum* hybrids. *Hereditas* **73**, 157–161.

Mettin, D., Blutner, W.D. and Schlegel, G. (1973) Additional evidence on spontaneous 1B/1R wheat-rye substitutions and translocations. Proc. 4th Int. Wheat Genetics Symp. Columbia, Mo. pp. 179–184.

Okamoto, M. (1957) Asynaptic effect of chromosome V. *Wheat Inf. Serv.* **5**, 6.

Riley, R. (1966a) The genetic regulation of meiotic behaviour in wheat and its relatives. Proc. 2nd Int. Wheat Genetics Symp., Lund, 1963. (*Hereditas* Suppl. Vol. 2) pp. 395–408.

Riley, R. (1966b) Cytogenetics and wheat breeding. *Contemp. Agric.* No. 11–12, 107–117.

Riley, R. and Chapman, V. (1958) Genetic control of the cytologically diploid behaviour of hexaploid wheat. *Nature* **182**, 713–115.

Riley, R., Unrau, J. and Chapman, V. (1958) The diploidisation of polyploid wheat. *Heredity* **15**, 407–429.

Riley, R., Chapman, V. and Johnson, R. (1968) The incorporation of alien disease resistance in wheat by genetic interference with the regulation of meiotic chromosome synapsis. *Genet. Res.* **12**, 199–219.

Sallee, P.J. and Kimber, G. (1979) An analysis of the pairing of wheat telocentric chromosomes. Proc. 5th Int. Wheat Genetics Symp., New Delhi, 1978. Vol. 1, pp. 408–419.

Sears, E.R. (1972a) Chromosome engineering in wheat. *Stadler Genet. Symp.* **4** 23–38.

Sears, E.R. (1972b) Reduced proximal crossing-over in telocentric chromosomes of wheat. *Genet. Iber.* **24**, 233–239.

Sears, E.R. (1977) An induced mutant with homoeologous pairing in wheat. *Can. J. Genet. Cytol* **19**, 585–593.

Sears, E.R. (1978) Analysis of wheat-Agropyron recombinant chromosomes. In 'Interspecific Hybridization in Plant Breeding'. (Proc. 8th Congr. Eucarpia). pp. 63–72.

Wall, A.M., Riley, R. and Chapman, V. (1971) Wheat mutants permitting homoeologous meiotic chromosome pairing. *Genet. Res.* **18**, 311–328.

Zeller, F.J. (1973) 1B/1R wheat-rye chromosome substitutions and translocations. Proc. 4th Int. Wheat Genetics Symp., Columbia, Mo. pp. 209–221.

6   PROSPECTS FOR WHEAT TRANSFORMATION

J. Langridge

Introduction

The capacity for transformation, that is, the uptake, maintenance and expression of foreign DNA in cells or organisms, has been demonstrated for numerous genera of bacteria, some blue-green algae, certain fungi and for cultured
animal cells. Of the major groups of organisms, flowering plants are exceptional
in that no reproducible means of transformation have yet been discovered. Such
a method of plant genetic manipulation, in addition to aiding experiments on
somatic cell genetics, might greatly enhance the scope of conventional plant
breeding. This chapter outlines attempts to construct vectors for the introduction of exogenous DNA into plant cells with particular reference to the
possibility of wheat transformation.

Experimental approach

Certain papers that have been published on plant transformation indicate
that foreign DNA is readily taken up by plant cells or protoplasts and that
some of the genes so introduced are transcribed and translated, although only
transiently (Uchimiya, 1979). These experiments suggest that stable and
reproducible transformation requires specifically constructed vectors. They
should contain DNA sequences providing origins of replication (replicators) or
sequences which permit high levels of vector integration into the plant genome,
and preferably sequences with both properties. In addition the vector must
contain a replicator for *Escherichia coli*, in which the molecular constructions
are carried out, and genes for the selection of bacterial transformants. In
practice we have used genes for resistance to the antibiotics kanamycin and trimethoprim, in the belief that they may serve for both bacterial and plant
transformation. These antibiotics are inhibitory to both classes of organism
and studies of *in vitro* transcription and translation, as well as transformations
in lower eukaryotes, suggest that many prokaryotic genes are expressible in
eukaryotic cells.

The range of potential replicators for plant cells is very limited. The possibilities comprise the origins of replication contained in organellar (chloroplast and
mitochondrial) DNAs, in the *Agrobacterium* tumour-inducing plasmid, in the

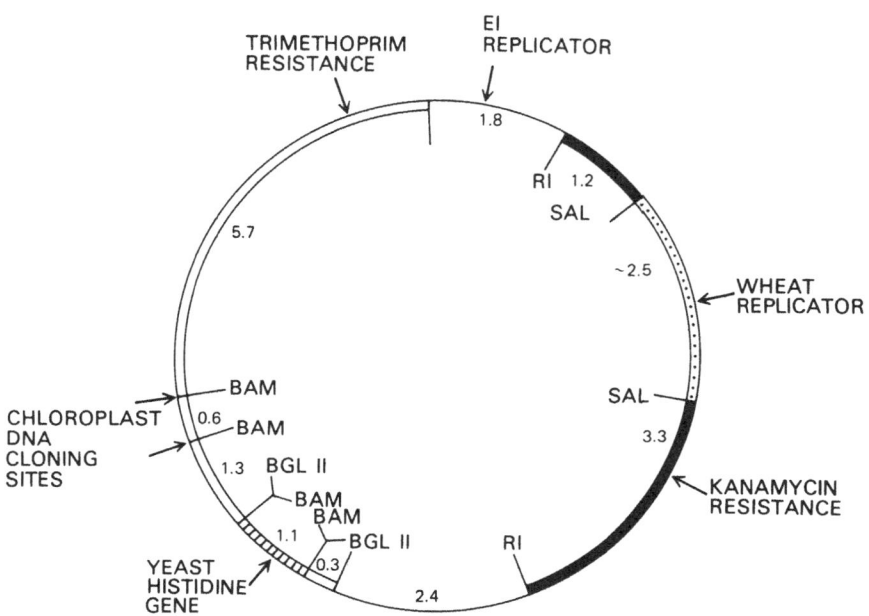

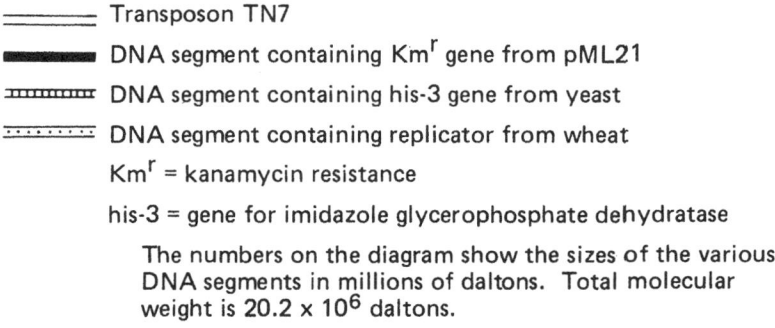

——— Col E1 plasmid

═══ Transposon TN7

▬▬▬ DNA segment containing Km^r gene from pML21

▭▭▭ DNA segment containing his-3 gene from yeast

⋯⋯⋯ DNA segment containing replicator from wheat

Km^r = kanamycin resistance

his-3 = gene for imidazole glycerophosphate dehydratase

The numbers on the diagram show the sizes of the various DNA segments in millions of daltons. Total molecular weight is 20.2×10^6 daltons.

Fig. 6.1. Vector for wheat transformation (to scale).

double-stranded DNA of cauliflower mosaic virus and in the yeast two-micron plasmid. The range is further limited in respect of wheat because monocotyledonous plants are not infected by *Agrobacterium* and wheat is not a host for cauliflower mosaic virus. Our earlier experiments have established that none of these DNA molecules is capable of multiplication in *E. coli*, but the advent of methods of yeast transformation (Hinnen *et al.*, 1978; Beggs, 1978) raises the possibility that replicators could be isolated in yeast. Accordingly, our present approach is to construct vectors containing the various DNA sequences designated above, to ligate them with random pieces of plant DNA and to use yeast as a host for the selection of replicators.

Construction of vectors

The structure of the vector expected to effect the transformation of wheat cells is shown in Figure 6.1. Its detailed construction will be described elsewhere but, briefly, it comprised the following steps. The gene for trimethoprim resistance was incorporated into the plasmid E1 by transposition to give a molecule of 12.9×10^6 daltons containing two Eco RI restriction sites. Distamycin A was used to protect the Eco RI site in the transposon and DNA containing the gene for kanamycin resistance was ligated into the Eco RI site in plasmid E1. Next, the his-3 gene of yeast was incorporated into the transposon by the ligation of Bam HI ends to those of Bgl II; the resultant hybrid restriction sites are resistant to both endonucleases. The vector, now of 17.7×10^6 daltons, has genes for expression in bacteria, presumptively in plants, and in yeast; it can replicate in *E. coli* but not in yeast or flowering plants. This molecule was ligated to wheat DNA cut with endonuclease Sal I, transformed to yeast lacking the his-3 gene and selection was made for histidine-independant colonies (Table 6.1).

The vector unligated to wheat DNA gave a few small colonies (less than 0.25 mm in diameter) which appeared to be due to abortive transformation. Thus the transforming DNA entered the cell and the gene for imidazole glycerophosphate dehydratase was expressed, but the incoming molecule failed to replicate. This class of colony also appeared when the vector was ligated to wheat DNA but, in addition, there were colonies of 1–2 mm and 4–7 mm in diameter. The smaller colonies were irregular in shape (Figure 6.2), their cells multiplied very slowly and they showed a low cloning efficiency. Such characteristics are associated with a frequent loss of the vector molecules. In contrast, the larger colonies were regular in outline, their cells grew at normal rates and they did not lose the his-3 gene. It has not been possible to reisolate vector from the stably transformed cells, but DNA extracts from the unstable ones have given a few colonies on transformation back to *E. coli*. The size of the inserted wheat DNA in one of these transformants is shown in the vector diagram of Figure 6.1.

Table 6.1

Results of transformations designed to isolate wheat chromosomal replicators.

Transforming DNA	his⁺ colonies
Nil	0
Vector alone (10 μg)	8 (small)
Vector (10 μg) + wheat DNA (10 μg)	68
Yeast episomal plasmid (3 μg)	52 690

The wheat DNA, kindly provided by Dr. E. Dennis, had its high-density satellite fraction removed and contained less than one percent of mitochondrial and chloroplast DNA. The yeast episomal plasmid, kindly provided by Dr. R.W. Davis, contained part of the two-micron yeast plasmid which allowed it to replicate in yeast cells.

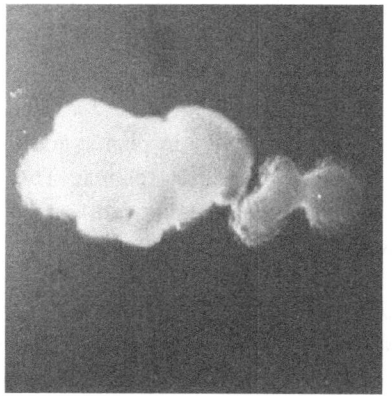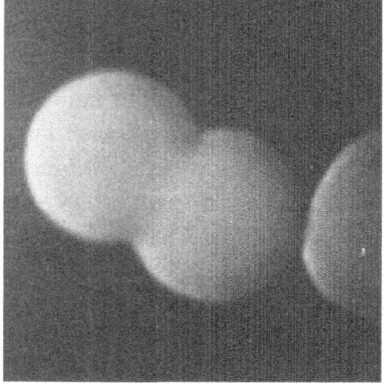

Fig. 6.2. Photographs of the two types of histidine-independent colonies obtained on transforming yeast with a ligation mixture of vector and wheat DNA. Left, unstable colonies; right, stable colonies.

Discussion

Transformation of yeast with vectors containing wheat DNA, as well as those with other plant DNAs, provides two types of histidine-independent colony, apart from the abortive transformants already mentioned. The unstable type is interpreted as containing a vector which includes a wheat DNA replicator. The instability may be due to the plant replicator having a low affinity for the yeast DNA polymerase or it may be due to competition with the replicators of the endogenous two-micron yeast plasmids. The stable type of transformant is interpreted as resulting from wheat DNA-facilitated integration of the vector, or at least the his-3 gene, into the yeast genome. Possible sources of the required sequence homology between wheat and yeast DNA include the ribosomal RNA genes, transfer RNA genes and histone genes among others unspecifiable. Although closer examination of the vectors will be required for reasonable certainty, we may have two types of wheat-transforming molecule. The first (the unstable type) is a replicative vector having only a low incidence of integration, and the second (the stable type) is an integrative vector which probably does not replicate in the yeast cell.

Before these vectors can actually be tested as transforming molecules, it will be necessary to develop means for growing isolated wheat cells in sterile culture, for forming protoplasts and for obtaining regeneration of actively-dividing cells from the protoplasts. This sequence of cultural events has not yet been obtained with wheat, but recent experiments with *Pennisetum* (Vasil and Vasil, 1980) suggest that it should be possible.

References

Beggs, J.D. (1978) Transformation of yeast by a replicating hybrid plasmid. *Nature* **275**, 104–109.

Hinnen, A., Hicks, J.B., and Fink, G.R. (1978) Transformation in yeast. *Proc. Nat. Acad. Sci. U.S.A.* **75**, 1929–1933.

Uchimiya, H. (1979) Progress in plant protoplast technology — isolation culture and genetic manipulation. *Rep. Inst. Agric. Res. Tohoku Univ.* **30**, 29–53.

Vasil, V., and Vasil, I.K. (1980) Isolation and culture of cereal protoplasts. Part 2. Embryogenesis and plantlet formation from protoplasts of *Pennisetum americanum. Theor. Appl. Genet.* **56**, 97–99.

7 NEW APPROACHES TO WHEAT BREEDING

C.J. Driscoll

Before considering new approaches to wheat breeding, consideration is given to the development of the types of breeding methods that are currently in use. Examples will be taken from Australian wheat breeding. The first wheats grown in Australia, which originated in England and Capetown, were unsuitable for Australian conditions (Waterhouse, 1936). The first improvements in the Australian wheats were brought about by farmers' selections. For example the selection of Purple Straw by Frame in 1860 and selection of rust-resistant wheats by Marshall from 1868 (Wrigley and Rathjen, 1981).

Further progress was achieved following deliberate hybridization. Farrer's wheat breeding, which began at Lambrigg near Canberra in 1886, included cross-breeding in that decade. Farrer realized the importance of the selection of parents for hybridization and the importance of selection procedures in ensuing generations. The earliest type of selection was pedigree selection and this persisted to beyond the mid-point of this century. There were difficulties with this method, viz., performance of populations of later-generation homozygotes had to be predicted from early generation heterozygotes and the method involved an inordinately-large amount of record-keeping.

The backcross method was introduced in 1942 by Pugsley with the advantage that useful traits could be introduced into adapted types without loss of favourable combinations of genes that had been previously constructed. Limited backcrossing became the preferred approach as this, at times, resulted in lines that performed better than the recurrent parent.

The next major discontinuity in breeding methodology was the development of the various forms of the progeny method, such as the F_2 progeny method adopted for wheat by Rathjen (Rathjen and Pugsley, 1977). This method, which was described by Lupton and Whitehouse (1957), involves classification of F_2 individuals by way of the yielding abilities of their progenies. This system requires a substantial infrastructure as the growing and analysis of a very large number of plots is essential. Large numbers are needed as selection is on the basis of grain yield and many selections must be subjected to replicated yield tests over a number of environments. In order to achieve this, specialized machinery for small plot trials had to be developed and the breeding programmes computerized.

Two reasons for the continued success of the wheat breeding methods adopted throughout the world are that they are not complicated to manage and they continually produce new parents for further hybridization (Williams, 1980). If a new method is to replace existing methods it must be easier to apply or result in an increased gain in yield. New approaches are not desperately needed as the current methods are successful and further advances are confidently expected (Bingham, 1980).

What is required is greater integration of studies in related sciences with wheat breeding. Cytogenetic or molecular genetic approaches to plant improve--ment will not replace plant breeding. These and other sciences will provide useful materials and information which will make the wheat breeder's role a more productive one. Plant breeding, as we know it today, is more than applied genetics (Mather, 1980). It involves a detailed knowledge of the industrial side of the growing and marketing of the product. It is the 'art section' of wheat breeding that geneticists and scientists of other disciplines must leave to the wheat breeder. Scientists of other disciplines can, however, have a part in this exciting field of achievement by strengthening the 'science section' of wheat breeding by presenting their scientific findings to the wheat breeder in a style that is compatible with the style of applied genetics used by the breeder.

Within this context new approaches to wheat breeding will be considered under four headings:
Increases in genetic variability for direct use or use in parental material;
Increases in combinations of alleles after hybridization;
Increases in selection efficiency;
Departures from homozygous, homogeneous varieties.

Increases in genetic variability

In addition to the collection and screening of natural variation and the induction of variation by mutagens, genetic variation has been increased by distant hybridization. The best example of this involves the transfer of wheat-stem-rust resistance gene *Sr26* from *Agropyron elongatum* (Host) Beauv. to chromosome 6A of wheat by Knott (1961). This alien resistance has been incorporated into Australian wheat varieties Eagle (Martin, 1971), Kite (Fisher and Martin, 1974), Jabiru (Fisher and Syme, 1977) and Avocet (Martin, pers. comm.).

Recent developments in this relatively new approach to wheat breeding include the widening of hybridization to crosses between wheat and barley (Shepherd and Islam, Chapter 8). Not only does this extend the crossing base beyond the sub-tribe *Triticinae*, but it involves crossing wheat to another established crop species.

The other major development in this area is the isolation of *ph* mutants by Riley (1968) and Sears (1977). Mutants of the 5BL gene will allow more precise transfer of alien segments to wheat (Sears, Chapter 5). This increase in precision will be valuable, as most transfers have been too gross to be easily incorporated into commercial varieties. It has also been pointed out by Shepherd (1977) that special efforts can be made to select genetic backgrounds that overcome the inherent disadvantage of low yield that is often associated with alien transfers. The method proposed involves the production of lines that are homozygous for the alien segment but heterozygous for many background genes. Selection for high yield in ensuing generations results in selection for accommodating genetic backgrounds. This is currently being attempted by Shepherd with transfers of segments of rye chromosome IR carrying wheat stem rust resistance, and by the author with the Transec translocation. The Transec translocation, which carries wheat-leaf-rust resistance and wheat-powdery-mildew resistance but is normally associated with low yield, involves transfer of a segment of rye chromosome 2R to chromosome arm $4A\beta$ (Driscoll and Bielig, 1968). This is regarded as a non-homoeologous transfer but this conclusion may need modification following the finding that rye chromosome 2R partially restores fertility to a male-sterility mutant located on chromosome 4A (see below).

Another way in which genetic variability has been increased is by way of anther culture (Hu Han *et al.*, 1980). This variation may almost entirely involve gross chromosome changes; however, greater claims for increasing variation are made by way of 'somaclonal variation' (Scowcroft, 1980) in other species, especially sugar cane and potatoes.

In parallel with these new methods of increasing genetic variability there have been meaningful developments in recognizing and determining the relationships of wheat and alien chromosomes. These include the use of *cRNA* in cytological hybridization (Peacock *et al.*, Chapter 3) and the considerable improvement in the application of N- and C-banding techniques in wheat (Gerlach, 1977; Jewell, 1979).

Increases in combinations of alleles

Most wheat-breeding programs involve hybridization of two homozygous or near-homozygous parents followed by several self pollinations. Selection is imposed during the serial selfings in a variety of ways. Many variations of this basic scheme exist, including the recycling of F_5 and F_6 materials as parents in further rounds of crosses and in some cases the intercrossing of F_1's. However, a general feature of the majority of breeding programs is rapid return to homozygosity and the concurrent rapid decrease in effective crossing over. Effective crossing over is defined as crossing over between two heterozygous loci. If one or both loci are homozygous, the recombinants are not different from the parents and the crossing over is ineffective.

A relatively new approach to wheat breeding is being evaluated at the Waite Agricultural Research Institute. This involves management of populations involving a male sterility mutant such that a greater proportion of crossovers are effective crossovers. Composite Crosses have been established based on the 'Cornerstone' male sterility mutant that involves a γ-ray-induced recessive mutation on chromosome arm $4A\alpha$ (Driscoll and Barlow, 1976; Driscoll, 1977). In Composite Cross I, Cornerstone was crossed to seven high-yielding varieties and the F_1-derived seed was mechanically mixed and planted in a grid pattern. Ten seeds were harvested from each of 100 male sterile plants and re-planted. This procedure was repeated in the subsequent two generations and fertile segregants (all of which were heterozygous *Ms1c/ms1c*) were selected. Homozygous fertile lines were isolated by progeny testing and these are being evaluated for yield in the 1980 season by Mr. I.D. Kaehne and the author. In essence, F_4 plots, after zero or three rounds of outcrossing, are being compared. Other composites are being established with aims other than yield *per se*. The first of these involves gibberellic acid (GA) insensitivity. The Cornerstone mutant *ms1c* was induced in the α arm of chromosome 4A of Pitic 62 which also bears *gai/rht1* on that arm (Gale and Law, 1977). Considering that *ms1c* has been mapped at least 50 units from the centromere (Barlow and Driscoll, 1980) and *Gai/Rht1* has been mapped 15±3 map units from the centromere (McVittie *et al.*, 1978), one would expect the *ms1c Gai/Rht1* recombinant would occur frequently in the appropriate segregating generation. However, so far this recombination has not been forthcoming. If it is not obtained in larger populations, the combination *ms1c Gai/Rht2* can presumably be obtained without difficulty since *Gai/Rht2* is on chromosome 4D (Gale *et al.*, 1975).

The purpose of combining male sterility with GA insensitivity is to construct a Composite Cross on Cornerstone and a group of varieties all of which carry the same GA-insensitivity allele. This will afford the opportunity of selecting for an accumulation of minor genes for height on a GA-insensitive background. This would increase the opportunity of selecting for 'tall dwarfs' as suggested by Gale and Law (1977). Intercrossing increases the chances of bringing together the various minor genes for height in the different varieties involved in the Composite. Selection will involve taking the tallest individuals of each generation with the realization that height increases are in spite of homozygosity for the chosen *Gai/Rht* allele.

Composite Crosses are also expected to be useful in cases where major genes for resistance to a particular disease are not known. Such a case exists with the

take-all disease of wheat caused by *Gaeumannomyces graminis* (Sacc.) v. Arx
& Olivier var. *tritici* Walker. Several wheat lines have been identified that
appear to carry minor degrees of resistance to take-all. It is proposed to establish a Composite Cross based on Cornerstone and involving these varieties and to
place the succeeding populations in *Gaeumannomyces*-infested soil. Artificial
infestation is possible using techniques established by Rovira (1977).

Planned and managed male sterility-based composites are different from
Evolution Breeding Composites as employed by Suneson *et al.* (1963) but
similar to those used by Ramage (1977) in barley, in that they are established
for distinct purposes and seed is taken from male-sterile plants only, during the
outcrossing phase. These composites may result in useful genotypes by way of
increasing the number of effective recombination events.

Increases in selection efficiencies

Distinctly different methodologies such as haploid breeding (Ouyang *et al.*,
1973) and single seed descent (Knott and Kumar, 1975) have been used in
wheat breeding in order to shorten the time needed to reach homozygosity.
However, of greater significance is the emphasis on selection for specific physiological traits. These include rather dramatic changes brough about by single
genes such as those governing daylength insensitivity (*ppd*), vernalization
response (*Vrn*) and gibberellic acid insensitivity/reduced height (*Gai/Rht*)
(Pugsley, 1979).

The *Gai/Rht* genes have had a dramatic effect on world agriculture. All
three genes *Gai/Rht1* and *Gai/Rht3* on chromosome 4A and *Gai/Rht2* on
chromosome 4D reduce height and render the plant insensitive to gibberellic
acid (Gale and Law, 1977). Other mutants affecting hormonal functions in
the plant may be identified, and these may play a part in changing the plant such
that it is more productive. Thus this attribute in the wider sense could be of
major significance in further yield increases.

Another relatively recent change which is of a very different nature, is the
massive scale of breeding operations initiated by the International Agricultural
Research Centres. CIMMYT, for example, makes a very large number of wheat
crosses each year and retains very large numbers of selections in the ensuing
generations. This increased size of the breeding program provides much more
diversity for selection in any given year. Selection is also more efficient
because of the increases in cooperative ventures which result in an accumulation of more information about genotypes. The International Wheat Yield
Nursery and the National Rust Control Programme in Australia are two such
examples.

Departures from homogeneous, homozygous varieties

Two new approaches which have received attention recently are multilines and hybrid varieties. Multilines, which are heterogeneous populations of homozygotes, are currently regarded as a possible means of combating virulence changes in pathogens (Jensen, 1952; Borlaug, 1959; Marshall, 1977). A multiline variety 'Tumult' has been released in the Netherlands for this purpose (Groenewegen, 1977). One type of multiline, viz. simple variety mixtures, would involve minimal breeding effort (Wolfe, 1978). The compatibility of simple variety mixtures with wheat quality requirements, and the interaction of such multilines with disease developments, should be ascertained with wheat bearing in mind the experiences with oats (Frey *et al.*, 1977) and barley (Wolfe and Barrett, 1980).

Hybrid wheat, based upon cytoplasmic male sterility derived from *Triticum timopheevi* Zhuk., went through a vigorous period of investigation in the 1960s and 1970s, (Schmidt *et al.*, 1962). However, the work has currently lost its momentum, mainly because of difficulties in fertility restoration. Restoration is not simply inherited and is considerably affected by environmental factors. This led to difficulties of realizing yield increases in hybrids as well as to production difficulties. Also, perhaps *T. timopheevi* cytoplasm *per se* depresses yield.

An alternative method of producing hybrid wheat is currently being researched at the Waite Institute. This system involves normal *aestivum* cytoplasm and a chromosomal male-sterility mutant (Driscoll, 1972). The Cornerstone male sterility-mutant on chromosome 4A is being used in this study and a search is underway for a chromosome that will restore fertility. Not only must this chromosome restore fertility but it must refrain from pairing at meiosis with any of the wheat chromosomes. Finding such a chromosome has not proved to be easy; however, three different possibilities are currently being analysed (Driscoll, 1980). The first involves barley chromosome 4H. Islam and Shepherd (Chapter 8) have substituted a structurally-modified chromosome 4H of barley for chromosome 4A and the resultant plant is male fertile. As to whether the normal 4H restores fertility is the subject of current investigation. Thus, the normal or the modified 4H is a possible chromosome for this purpose.

The second involves a chromosome or chromosomes of cereal rye (*Secale cereale* L.). Cornerstone has been crossed with rye and the chromosomes of the F_1 doubled so as to produce an amphiploid. This amphiploid is fertile (Hossain and Driscoll, 1980); thus the rye genome is capable of restoring fertility. The BC_1 to male-sterile hexaploid wheat is also fertile, and this has been used to produce a BC_2 population, part of which has been analysed. It is of considerable interest that a 43-chromosome segregant, homozygous *ms1c/ms1c* and containing 2R, is partly male fertile. This indicates that exchange between homoeologous groups 2 and 4 has occurred in either wheat or rye. Perhaps another rye chromosome also partly contributes to fertility restoration. If two chromosomes are involved, centric fusion (Sears, 1972) could perhaps reduce these to one.

The third possibility involves *Triticum monococcum* L. The F_1 of male-sterile *T. durum* Desf. homozygous *ms1c/ms1c* (Hossain and Driscoll, 1980) by *T. monococcum*, treated with colchicine, produced a fertile AAAABB amphiploid; thus the *monococcum* genome is capable of restoring fertility, which is not surprising. What is significant is that in the undoubled F_1, which contains AAB genomes, chromosome 4A of *monococcum* fails to pair with 4A of *durum*. Thus if *monococcum* 4A carries full fertility restoration it may prove to be a suitable chromosome for this system.

If production of hybrids by this system becomes genetically possible, it offers good opportunities for gene management in terms of disease resistance. This is particularly so because the male parent of the hybrid is totally unaltered in terms of fertility/sterility components. Therefore male parents with different disease resistance genes could be strategically deployed to combat changes in virulence patterns.

References

Barlow, K.K. and Driscoll, C.J. (1980) Linkage studies involving two chromosomal male-sterility mutants in hexaploid wheat. *Genetics* (in press).

Bingham, J. (1980) Achievements of conventional breeding. In 'The Manipulation of Genetic Systems in Plant Breeding'. The Royal Society Discussion Meeting, London, 1980, (in press).

Borlaug, N.E. (1959) The use of multilineal or composite varieties to control airborne epidemic diseases of self-pollinated crop plants. Proc. lst. Int. Wheat Genetics Symp., Winnipeg, 1959. pp. 12–26.

Driscoll, C.J. (1972) XYZ system of producing hybrid wheat. *Crop Sci.* **12**, 516–517.

Driscoll, C.J. (1977) Registration of Cornerstone male-sterile wheat germplasm. *Crop Sci.* **17**, 190.

Driscoll, C.J. (1980) Perspectives in chromosome manipulation. In 'The Manipulation of Genetic Systems in Plant Breeding.' The Royal Society Discussion Meeting, London, 1980, (in press).

Driscoll, C.J. and Barlow, K.K. (1976) Male sterility in plants: Induction, isolation and utilization. In 'Induced Mutations in Cross-breeding.' IAEA, Vienna. pp. 123–131.

Driscoll, C.J. and Bielig, L.M. (1968) Mapping of the Transec wheat-rye translocation. *Can. J. Genet. Cytol.* **10**, 421–425.

Fisher, J.A. and Martin, R. (1974) Kite, Condor and Egret — three new wheats. *Agric. Gaz. N.S.W.* **85**, 10–13.

Fisher, J.A. and Syme, J.R. (1977) Register of Cereal and Linum Varieties in Australia. Wheat. Jabiru. *J. Aust. Inst. Agric. Sci.* **43**, 173.

Frey, K.J., Browning, J.A. and Simons, M.D. (1977) Management systems for host genes to control disease loss. *Ann. N. Y. Acad. Sci.* **287**, 255–274.

Gale, M.D. and Law, C.N. (1977) Norin-10-based semidwarfism. In 'Genetic Diversity in Plants.' (Eds. A. Muhammed, R. Aksel and R.C. von Borstel.) Plenum Press, New York and London.

Gale, M.D., Law, C.N. and Worland, A.J. (1975) The chromosomal location of a major dwarfing gene from Norin 10 in new British semi-dwarf wheats. *Heredity* **35**, 417–421.

Gerlach, W.C. (1977) N-banded karyotypes of wheat species. *Chromosoma (Berl.)* **62**, 49–56.

Groenewegen, L.J.M. (1977) Multilines as a tool in breeding for reliable yields. *Cereal Res. Commun.* **5**, 125–132.

Hossain, M.A. and C.J. Driscoll. (1980) Transfer of Cornerstone male sterility from hexaploid wheat to tetraploid wheat and incorporation into hexaploid and octoploid triticales. *Can. J. Genet. Cytol.* (in press).

Hu Han, Xi Ziying, Oyang Junwen, Hau Shui, He Mengyan, Xu Zongyao and Zon Mingqian. (1980) Chromosome variation of pollen mother cell of pollen-derived plants in wheat (*Triticum aestivum* L.) *Sci. Sin.* **23**, 905–914.

Jensen, N.F. (1952) Intravarietal diversification in oat breeding. *Agron. J.* **44**, 30–34.

Jewell, D.C. (1979) Chromosome banding in *Triticum aestivum* cv. Chinese Spring and *Aegilops variabilis. Chromosoma (Berl.)* **71**, 129–134.

Knott, D.R. (1961) The inheritance of rust resistance. VI. The transfer of stem rust resistance from *Agropyron elongatum* to common wheat. *Can. J. Plant Sci.* **41**, 109–123.

Lupton, F.G.H. and Whitehouse, R.N.H. (1957) Studies on the breeding of self-pollinating cereals. I. Selection methods in breeding for yield. *Euphytica* **6**, 169–184.

Marshall, D.R. (1977) The advantages and hazards of genetic homogeneity. *Ann. N. Y. Acad. Sci.* **287**, 1–20.

Martin, R.H. (1971) Eagle — a new wheat cultivar. *Agric. Gaz. N.S.W.* **82**, 207.

Mather, K. (1980) The prospects. In 'The Manipulation of Genetic Systems in Plant Breeding.' The Royal Society Discussion Meeting, London, 1980, (in press).

McVittie, J.A., Gale, M.D., Marshall, G.A., and Westcott, B. (1978) The intra-chromosomal mapping of the Norin 10 and Tom Thumb dwarfing genes. *Heredity* **40**, 67–70.

Ouyang, T.W., Hu, H. Chuang, C.C. and Tseng, C.C. (1973) Induction of pollen plants from anthers of *Triticum aestivum* L. cultured *in vitro*. *Sci. Sin.* **16**, 79–95.

Pugsley, A.T. (1979) Semi-dwarf wheats and gibberellin. *Annual Wheat Newsletter* **25**, 42.

Ramage, R.T. (1977) Male sterile facilitated recurrent selection and happy homes. Proc. 4th Regional Winter Cereal Workshop, Barley, Amman, April 1977.

Rathjen, A.J. and Pugsley, A.T. (1977) Wheat breeding at the Waite Institute, 1925–1978. In 'Biennial Report of the Waite Agricultural Research Institute 1976–77.' Adelaide. pp. 12–37.

Riley, R. (1968) The basic and applied genetics of chromosome pairing. Proc. 3rd Int. Wheat Genetics Symp., Canberra, 1968. pp. 185–195.

Rovira, A.D. (1977) Manipulation of the level of take-all disease in the field by inoculation. Proc. Australian Plant Path. Soc. Workshop on Epidemiology and Crop Loss Assessment, Lincoln College, N.Z. 1977. pp. 11-1 to 11-4.

Schmidt, J.W., Johnson, V.A. and Mann, S.S. (1962) Hybrid Wheat. *Nebr. Exp. Stn. Q.* **9**, 9.

Scowcroft, W.R. (1980) Tissue culture research: Current status and future potential. Rice Tissue Culture Planning Conference, IRRI, Philippines, April, 1980.

Sears, E.R. (1972) Chromosome engineering in wheat. *Stadler Genet. Symp.* **4**, 23–38.

Sears, E.R. (1977) An induced mutant with homoeologous pairing in common wheat. *Can. J. Genet. Cytol.* **19**, 585–593.

Shepherd, K.W. (1977) Utilization of a rye chromosome arm in wheat breeding. 3rd Intern. Cong. of SABRAO, Canberra, February, 1977. Vol. 2(a) 16–20.

Suneson, C.A., Pope, W.K., Jensen, N.F., Poehlman, J.M. and Smith, G.S. (1963) Wheat Composite Cross I. Created for breeders everywhere. *Crop Sci.* **3**, 101–102.

Waterhouse, W.L. (1936) Some observations on cereal rust problems in Australia. *Proc. Linn. Soc., N.S.W.* **61**, v–xxxviii.

Williams, W. (1980) Methods of production of new varieties. In 'The Manipulation of Genetic Systems in Plant Breeding.' The Royal Society Discussion Meeting, London, 1980, (in press).

Wolfe, M.W. (1978) Some practical implications of the use of cereal variety mixtures. In 'Plant Disease Epidemiology.' (Eds. P.R. Scott and A. Bainbridge.) Blackwell Scientific Pub., Oxford. pp. 201–207.

Wolfe, M.S. and Barrett, J.A. (1980) Can we lead the pathogen astray? *Plant Disease* **64,** 148–155.

Wrigley, C.W. and Rathjen, A.J. (1981) The contribution of wheat breeding to the success of wheat growing in Australia. In 'People and Plants in Australia: Contributions towards a History of Australian Botany.' (Eds. D.J. Carr and S.G.M. Carr.) Academic Press (in press).

8 WHEAT:BARLEY HYBRIDS – THE FIRST EIGHTY YEARS

K.W. Shepherd and A.K.M.R. Islam

Beginning with William Farrer in 1896, many cereal breeders around the world have shown interest in hybridizing wheat and barley, the two most important cereals of temperate agriculture. Although the aims in attempting this wide cross usually were not stated, we can assume that the breeders were hoping to produce a new type of crop plant which combined desirable characteristics from both parent species. However, there is considerable doubt as to whether any true hybrids were produced in any of this early work. Recently Kruse (1973) provided good evidence that he had been successful in hybridizing wheat and barley, and there is now renewed interest in this subject, with much progress being achieved in the last seven years.

Early attempts at production of hybrids

Apparently Farrer (1904) was the first to publish on wheat:barley hybridization, reporting that on the 21st October 1896, he attempted "an unusually interesting cross" involving the transfer of pollen from Nepaul barley onto 12 emasculated florets of an early maturing variant of Blount's Lambrigg wheat. One small shrivelled seed was obtained which produced a self-fertile plant resembling the wheat parent except for its lighter green leaves and weaker straw.

In the F_2 generation Farrer was surprised to find that the plants showed only minor morphological differences from the original wheat parent. The F_2 plant of 'best' appearance was selected and its seeds sown in a row. These F_3 plants were very uniform in growth and morphology but were recognizably different from the wheat parent and Farrer gave them the varietal name Bobs. Farrer remained puzzled by these observations until 1903 when some additional results were obtained from crosses made by his assistant, J.T. Pridham, between some rust-resistant Fife varieties and the same Nepaul barley. When Fife was used as the female parent the crosses were unsuccessful but seven small shrivelled seeds were obtained on the barley parent in the reciprocal cross. The juvenile leaves of most of these presumptive hybrid plants were distinctly different from those of the barley parent and Farrer was optimistic that these plants represented true hybrids. However, as the plants matured the differences from Nepaul barley became less and less obvious.

Farrer reasoned that if Bobs had barley parentage it might be more easily crossed with barley pollen than other wheats and he arranged for his assistants, Pridham at Canberra and Hurst at Wagga, to make the crosses with Nepaul barley. Although 23 seeds were obtained and sown in 1904 no information is available on the morphology and breeding behaviour of these plants.

Since Farrer was unable to demonstrate unequivocally the presence of any barley characters in the presumptive wheat x barley hybrids and their progeny nor any wheat characters in the presumptive barley x wheat hybrids, he remained uncertain of whether any of these plants represented true hybrids (Macindoe, 1939).

Another wheat variety, Canberra, was claimed by Pridham (1914) to have come from a cross between Federation wheat as the female parent and Volga barley. Later, Waterhouse (1930) attempted to repeat these crosses, using both Nepaul and Volga, and several wheat cultivars, including Federation. Only five seeds were obtained from 961 pollinated wheat florets. Since these plants all resembled the wheat parent he considered that they were from uncontrolled self-fertilization. Pye, the prominent cereal breeder at Dookie Agricultural College, Victoria, also attempted wheat:barley matings. Although a few seeds were obtained, he concluded that the plants were probably not hybrids (Pye, 1921, cited by Sims, 1980).

Gordon and Raw (1932) have described the wheat:barley hybridizations attempted at the State Research Farm, Werribee, Victoria, from 1913 to 1932. They concluded that none of their plants were true wheat:barley hybrids, but were convinced that these seeds had not come from uncontrolled out-crosses; instead, because of the consistent occurrence of irregular segregation ratios for various characters in the F_2 and F_3 progeny, and some cytological irregularities observed in the progeny of one of these crosses, they concluded that there was a disturbance of the chromosome constitution of the ovule parent. However, in our opinion most, if not all, of their results are consistent with uncontrolled outcrossing.

More recently Ahokas (1970) reported some further unsuccessful attempts at making wheat:barley hybrids. He used several novel treatments such as mixing fresh barley pollen with inactivated wheat pollen, and applying gibberellic acid and RNase to the stigma before and after pollination, but no hybrids were obtained.

Thus, there is no good evidence that any of the putative hybrids reported in the early papers represent crosses between wheat and barley.

Recent production of wheat:barley hybrids

Kruse (1973) provided the first good evidence that wheat:barley hybrids could be produced. He pollinated hexaploid wheat (*Triticum aestivum*) with diploid barley (*Hordeum vulgare*) and obtained a few seeds, but their embryos died in culture. However, he produced hybrid plants with the reciprocal cross. Kruse pollinated diploid barley with diploid (*T. monococcum*), tetraploid (*T. dicoccum*) and hexaploid wheat, and obtained seed sets of 0.25%, 1% and 3%, respectively. Approximately 90% of the seeds contained embryos, and when these were cultured, some hybrid plants were obtained for each cross combination. The somatic chromosome numbers of the hybrids using *T. monococcum, T. dicoccum* and *T. aestivum* as the pollen parent were 14, 21 and 28, respectively, as expected. The satellite chromosomes of both wheat and barley were identified in the hybrids, and in each cross the F_1 hybrids resembled the wheat parent. The hybrids grew vigorously but were self-sterile and, initially, Kruse obtained only one plant after back-crossing the *H. vulgare* x *T. aestivum* hybrid with pollen from *T. aestivum*. Kruse (1976) reported an amphiploid after treating a barley x hexaploid wheat hybrid with nitrous oxide, but this plant remained self-sterile.

Following Kruse's results, Islam in 1973 initiated a program of wheat and barley crosses using a wide range of species and cultivars. The barley cultivars included Ketch, Clipper, Prior and Betzes. The wheat parents included the hexaploids, Chinese Spring and four Australian cultivars (Gabo, Falcon, Heron and Halberd) and two tetraploid wheats, *T. dicoccum* and *T. durum*. The crosses were made, initially using Kruse's (1973) crossing procedure, but it was found that hybrid seeds could be produced just as easily on the intact plant, and excision of spikes was discontinued.

No seeds were obtained when Chinese Spring was pollinated with Ketch, but when barley was used as the ovule parent, seeds were obtained in most cross combinations (Table 8.1). Two hybrid plants with 21 chromosomes were obtained from Ketch x *T. dicoccum*, and one from Betzes x *T. dicoccum*. They were self-sterile and no backcross progeny were obtained. The barley x *T. durum* crosses yielded many large seeds but the derived plants died as seedlings.

The average seed set for the crosses with hexaploid wheats was 5.8%, with Betzes x Chinese Spring being the most successful combination giving 15.4% seed set (Islam *et al.*, 1975). Altogether 137 embryos were taken from the 211 seeds produced and 67 survived as viable plants. All plants were treated with colchicine, but no amphiploids were obtained. However, these plants were female fertile and gave 0.3 to 1.2 first backcross (BC_1) seeds/spike when pollinated with wheat.

The morphology (Figure 8.1), chromosome constitution and sterility of these

Table 8.1

Barley x wheat crosses. Number of florets pollinated and percentage seed set.

Wheat (pollen) parent / Barley (ovule) parent	Tetraploids		Hexaploids				
	T. dicoccum	*T. durum*	Chinese	Gabo	Heron	Falcon	Halberd
Ketch	195(3.0)*	42(16.7)	512	421	140	59	176
			(10.9)	(3.1)	(1.4)	(6.8)	(1.1)
Clipper	506(2.7)	–	576	191	53	65	82
			(9.7)	(2.6)	(0)	(3.1)	(0)
Prior	861(0.5)	30(10.0)	390	341	219	54	145
			(10.8)	(0.3)	(0.5)	(0)	(0)
Betzes	47(4.2)	33(18.2)	162	22	–	–	43
			(15.4)	(9.1)			(0)

( )* = % seed set.

Table 8.2

Wheat x barley crosses. Number of florets pollinated and percentage seed set.

Barley (pollen) parent / Wheat (ovule) parent	Betzes	Clipper	Ketch	Golden Promise
T. aestivum				
Chinese Spring	3381(1.3)*	902(0.3)	2731(0.2)	
Gabo	759(0)	182(0.5)		
Tobari 66	178(0)			
T. durum				
Kingfisher	817(1.6)			
K 713	1194(2.4)			337(3.8)
K 720	226(0)			

( )* = % seed set.

K   F₁   CS
          K x CS

Fig. 8.1. Spike morphology of the parents and the barley x wheat F₁ hybrid (x0.5). K = Ketch barley; CS = Chinese Spring wheat. (From Islam *et al.*, 1975).

plants were consistent with Kruse's observations on barley x hexaploid wheat hybrids, except for some chromosome number mosaicism in pollen mother cells (PMCs). The majority of plants exhibited partial pistillody, where a variable number of florets had one or more stamens replaced by pistil-like structures. The pistillody, which appears to be caused by an incompatibility between the wheat genome and barley cytoplasm, became more pronounced in later generations and severely interfered with attempts to produce alloplasmic wheat-barley addition lines. Efforts were made over the period 1975 to 1978 to produce the reciprocal cross so that the F_1 hybrids and derivatives would have the cytoplasm of wheat instead of barley (Islam *et al.*, 1978, 1981). The best result again was obtained with Chinese Spring x Betzes (Table 8.2), but the success rate was only 1.3% compared with 15.4% in the reciprocal cross. Forty-seven seeds were obtained and 22 possessed culturable embryos which yielded 20 plants. Only one of these was a normal F_1 hybrid having 21 wheat chromosomes and 7 barley chromosomes, forming 28 univalents at meiosis. The other 19 hybrids possessed somatic chromosome numbers ranging from 21 to 36, and included three plants which were wheat haploids, five plants which possessed a haploid complement of wheat chromosomes plus 1 to 6 different barley chromosomes, and eleven plants which had abnormal chromosome constitutions (Islam *et al.*, 1981). Islam and Shepherd (1981a) argued that the abnormal chromosome constitution of these wheat x barley hybrids could be due to the disruption of normal chromosome disjunction during early zygotic divisions, sometimes resulting in complete elimination of the barley chromosomes, and to duplication and deficiency of some wheat and barley chromosomes.

The morphology of the F_1 hybrids was also variable (Figure 8.2). The normal 28-chromosome hybrid resembled the reciprocal hybrid in morphology (Figure 8.2a), cytology and self-sterility but there was no evidence of pistillody. Most PMCs possessed 28-chromosomes but some chromosome number mosaicism was observed, as in the reciprocal hybrid (Islam and Shepherd, 1981a). In a few of the 28-chromosome PMCs there were one or more bivalents with the average chromosome association being 26.51 ′ + 0.72 ″ + 0.015 ‴ in 128 cells. This hybrid plant was treated with colchicine and although doubled sectors with 56 chromosomes were observed at meiosis, no seeds were produced from selfing. An average of 3.5 seeds/spike were produced by backcrossing with hexaploid wheat pollen.

The 23- and 21-chromosome plants resembled Chinese Spring haploids, except that the former plant had larger florets (Figure 8.2d,e). The spikes of the F_1 hybrid plants with very abnormal chromosome constitution exhibited numerous spike abnormalities, including increased awning, supernumerary spikelets and malformed florets (Figure 8.2b,c). All these F_1 hybrids were self-sterile, and some were female sterile. The normal F_1 hybrid and three of the abnormal hybrids have been used in the production of addition lines.

Recently these wheat x barley crosses were extended to include some tetra-ploid wheats and a few seeds were produced (Table 8.2). Seventeen plants died at the two-leaf stage within 14 days of their transfer from the agar culture

Fig. 8.2. Spike morphology of the wheat parent and some wheat x barley F_1 hybrids. (x0.75). CS = Chinese Spring parent; a = 28-chromosome normal hybrid; b = 31-chromosome hybrid; c = 27-chromosome hybrid; d = 23-chromosome hybrid; e = 21-chromosome hybrid (haploid wheat). (From Islam *et al.*, 1981).

medium to soil; two plants which survived and grew vigorously were haploids of *T. durum*.

A parallel program of crosses between Betzes barley and Chinese Spring wheat (Fedak, 1977, 1980) has produced reciprocal hybrids with frequencies of 0.80% of pollinated florets for the barley x wheat crosses and 0.23% for the reciprocal cross. These compare with 3.7% and 0.56% in our study. Whereas we observed a marked difference in the somatic chromosome constitution of the reciprocal hybrids, Fedak obtained mainly 28-chromosome hybrids in both types of crosses. This could be due to differences in the success rate of culturing embryos, since Fedak obtained only 3% success (5/152), whereas we obtained 91% success (20/22). It is possible that only embryos with the normal complement of 28 chromosomes survived in Fedak's (1980) study whereas we recovered a representative sample of all the embryo types produced. Another possibility is that the growth conditions of the parental plants influenced the outcome of these crosses. In Fedak's study the parental plants were raised in a controlled environment cabinet at 16 hours photoperiod, whereas we grew our plants under normal glasshouse conditions in spring and early summer with day lengths varying from 11 to 14 hours.

In both studies it was noted that the PMCs of the hybrids possessed variable chromosome numbers, but Fedak observed hyperploids only. This chromosome mosaicism was attributed to abnormal premeiotic mitoses in archespore cells (Fedak, 1980; Islam and Shepherd, 1981a). Fedak observed a higher level of chromosome pairing than we did at metaphase I in the 28-chromosome PMCs of both barley x wheat (1.72 "vs. 0.78 ") and the reciprocal hybrids (1.21 "vs. 0.72 "). Because the level of pairing observed in the hybrid was greater than the sum of that occurring in wheat and barley haploids, Fedak (1977, 1980) suggested that the barley genome may act as a weak suppressor of the Ph locus on chromosome 5B of wheat. We believe the evidence for this is not conclusive; some of the 28-chromosome PMCs could be aneuploid with duplication and deficiency of one or more chromosomes, so that the increased pairing could involve homologous rather than homoeologous chromosomes.

Bates *et al.* (1974) also claimed to have produced hybrids between durum wheat and barley, and bread wheat and barley, with the aid of immunosuppressant drugs injected into the female parent before pollination. The somatic chromosome number in root tips of the presumptive hybrids varied from 18-21 in durum crosses and 21-36 in bread wheat crosses. Since the supposed hybrids were self-fertile, they may have been aneuploid forms of wheat arising from mitotic abnormalities induced in embryonic cells by the immunosuppressant drugs. Bates *et al.* (1976) stated that wheat:barley hybrids obtained after treatment with immunosuppressants lose the barley chromosomes early in embryogenesis. These workers (Thomas *et al.*, 1977) later produced F_1 hybrids from *H. vulgare* x *T. turgidum* and *H. vulgare* x *T. aestivum* crosses without applying pre- or post-pollination chemical treatments. However, the rate of success was low except for crosses between Manker barley and *turgidum* and *aestivum* wheats, where seed sets of 1.7% and 2.5% were obtained. Only 7 plants were obtained from 32 cultured embryos. The hybrids had 21 and 28 chromosomes, respectively, although some mitotic and meiotic instabilities were noted. All were self-sterile and attempts to induce seed set by colchicine doubling were unsuccessful.

Besides these examples of hybridization between cultivated species of *Triticum* and *Hordeum*, crosses have been made between *Triticum* and wild species of *Hordeum*. Macindoe and Brown (1968) reported that Halloran had produced a wheat x wild barley hybrid, but details of this work have not been published. Apparently Halloran (pers. comm.) produced a 28-chromosome F_1 hybrid from Chinese Spring x *H. spontaneum* crosses without the aid of embryo culture. This hybrid plant was intermediate in phenotype to its parents but it was self-sterile and did not produce any seeds when backcrossed with wheat pollen.

Barclay (1975) crossed Chinese Spring with diploid and tetraploid strains of *H. bulbosum* but the derived plants were found to be 21-chromosome wheat haploids. Kimber and Sallee (1976) obtained one seed after pollinating 40 florets of *T. timopheevi* with *H. bogdanii*. The seed matured on the plant and germinated normally to produce a self-sterile hybrid with 21 somatic chromosomes. Martin and Chapman (1977) succeeded in producing a hybrid between *H. chilense* and Chinese Spring using embryo culture. The 28-chromosome hybrid was self-sterile, but a fertile amphiploid was produced after colchicine treatment (Chapman and Miller, 1978). The amphiploid has been used to produce heptaploids with 49 chromosomes and these are being used to produce addition lines having individual *chilense* chromosomes added to wheat.

To summarize, it appears that although numerous early attempts to hybridize wheat and barley were unsuccessful, after Kruse succeeded in producing hybrids between these two species in 1973, several other workers have successfully crossed various wheats, not only with *H. vulgare* but also with some wild *Hordeum* species including *H. spontaneum, H. bogdanii, H. bulbosum* and *H. chilense*.

Production, characterization and utilization of addition lines

Our main aims were to cross wheat and barley and then to produce the amphiploid, which in turn could be used to produce addition lines having individual pairs of barley chromosomes added to the chromosome complement of wheat. It was anticipated that these addition lines would be useful for assigning barley genes to particular chromosomes, determining the evolutionary relationship of wheat and barley chromosomes, and possibly for transferring desirable characters from barley to wheat. All attempts at producing fertile amphiploids from barley x wheat and wheat x barley hybrids failed; instead some 49-chromosome progeny (heptaploids) were obtained from both hybrids after backcrossing them with wheat pollen. These heptaploids, containing the full complement of wheat chromosomes and a haploid set of barley chromosomes, are of key importance in the production of addition lines (O'Mara, 1940) and we were able to proceed with our second aim.

Alloplasmic additions (barley cytoplasm)

Our first attempts at producing wheat-barley addition lines were made with the heptaploids containing barley cytoplasm (Islam *et al.*, 1975). These heptaploids exhibited a maximum chromosome association of 21 " + 7 ' at meiosis, indicating that they had come from pollination of egg cells containing a 28-chromosome restitution nucleus (Islam and Shepherd, 1980). The BC_1 plants were self-sterile and exhibited more pronounced pistillody than the F_1 hybrids, but a few BC_2 seeds were obtained after a second backcrossing with wheat pollen. Root tips from 62 of these BC_2 plants were screened cytologically and 8 putative monosomic addition lines with 43 chromosomes were obtained. However, the chromosome constitution of several of these plants could not be determined because of extreme pistillody and the absence of anthers. At least one of the plants was a monosomic addition, since it regularly formed 21 " + 1 ' at metaphase I but because of its self-sterility a disomic addition line could not be produced from it.

The widespread occurrence of pistillody in barley x wheat material, due to an incompatibility between the cytoplasm of barley and the wheat genome, has frustrated our attempts to produce alloplasmic wheat-barley addition lines, so efforts were then made to produce the additions in the cytoplasm of wheat instead of barley.

Euplasmic addition lines (wheat cytoplasm)

Production of lines — The first disomic addition line came from the 23-chromosome abnormal F_1 hybrid which exhibited 21 ' + 1 " at meiosis. When this self-sterile hybrid plant was pollinated with wheat the majority of the progeny seeds possessed 44 chromosomes which formed 22 " at meiosis. These plants were found to have a pair of barley chromosomes added to the chromosome complement of wheat. The 23-chromosome hybrid must have had the haploid complement of wheat plus a homologous pair of barley chromosomes. A restituted egg cell, with 23 chromosomes, gave the addition line directly when pollinated with wheat. Later, it was found that the barley chromosome present in this line was distinctly heterobrachial, unlike any of the Betzes chromosomes. We now suspect that this barley chromosome may possess a pericentric inversion with break points near the centromere of one arm and near the end of the other arm (Islam *et al.*, 1981). Two other addition lines involving the unmodified form of this barley chromosome were obtained, one from the 22-chromosome F_1 hybrid and the other from the 28-chromosome hybrid which formed 25 ' + 1 "' at meiosis (Islam *et al.*, 1981).

The other addition lines have come from the normal 28-chromosome hybrid. Heptaploids with 49 chromosomes were obtained in the BC_1 generation and a large population of BC_2 seeds was produced and screened cytologically for the presence of 43-chromosome individuals representing possible monosomic addition lines. A total of 25 monosomic additions which formed 21 " + 1 ' at meiosis were identified among 240 BC_2 plants and these, together with a few double-monosomic additions (21 " + 1 ' + 1 '), have been used to produce the disomic addition lines.

The 25 monosomic plants could be grouped into five different morphological classes, assumed to correspond to the addition of five different barley chromosomes to wheat. The progeny from one or two plants in each of these groups were first screened cytologically to detect plants with 44 somatic chromosomes and which formed 22 " at meiosis i.e. disomic additions. Only five of 789 progeny plants screened were of this type, but eight other plants were detected which formed 21 " + 1t " at meiosis. These monotelo-disomic additions have proved to be particularly useful since they gave rise to both disomic and ditelosomic additions with high frequency. The double monosomic additions have an advantage over the single monosomic additions, because they gave a higher yield of disomic additions in their progeny (2.0% vs. 0.6%) and also provide the opportunity for obtaining two different addition lines from the one stock.

We have investigated a new method of producing disomic additions, which does not depend on rare transmission of the extra barley chromosome through the pollen of the monosomic addition. This approach is based on the finding that wheat haploids can be obtained with high frequency when hexaploid wheat is hybridized with *H. bulbosum*, because the *bulbosum* chromosomes are eliminated during early divisions of the zygote (Barclay, 1975). Monosomic, mono-telo-disomic and disomic barley additions were crossed with *H. bulbosum* and 22-chromosome aneuhaploid progeny were obtained in high frequency in each case and disomic additions were obtained by doubling their chromosome number with colchicine (Islam and Shepherd, 1981b).

We have been able to produce six of the seven possible disomic additions, (designated A, B, C, D, E and F) and seven of the 14 ditelosomic additions of barley chromosomes to wheat. The seventh barley chromosome (G) could not be obtained as a disomic addition, because plants carrying this chromosome in single dose exhibit cytological abnormalities during microsporogenesis and are self-sterile.

Characterization of lines — The authenticity and individuality of the six presumptive disomic addition lines was checked by determining the chromosome pairing patterns at meiosis in progeny obtained from crossing each line with wheat (21 " + 1 '), and from intercrossing the lines to complete a half diallel (21 " + 1 ' + 1 ').

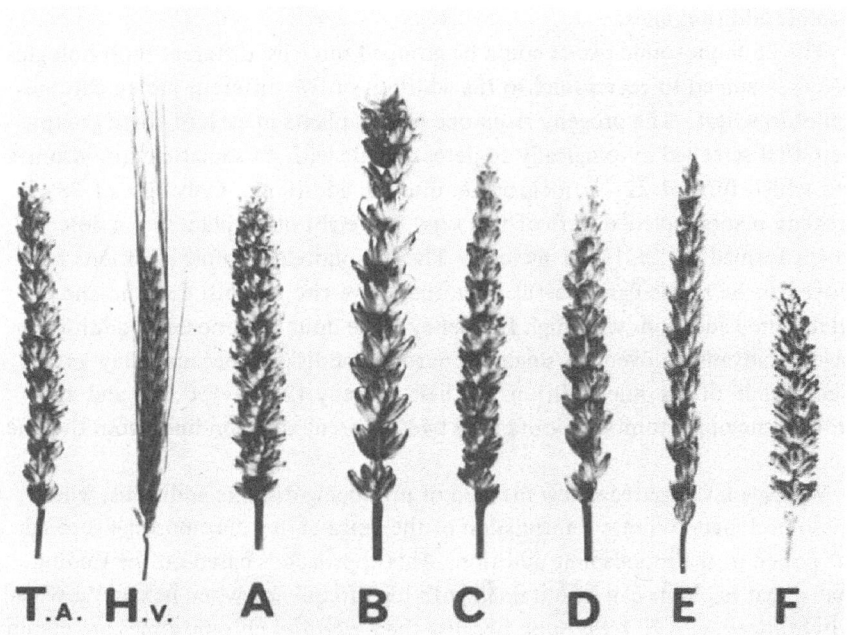

Fig. 8.3. Spike morphology of Chinese Spring wheat (T.A.), Betzes barley (H.V.), Disomic additions A,B,C,D,E and F. (x0.5). (From Islam *et al.*, 1981).

The lines were first characterized by their morphological and physiological differences from the Chinese Spring parent and in general the features which distinguished the monosomics were accentuated in the disomics. The spike characters of these lines are shown in Figure 8.3.

Whilst the addition lines are morphologically different from the wheat parent, they do not exhibit any obvious characters of barley, and therefore gross morphology has no analytical value for indicating which specific barley chromosome has been added to wheat. Biochemical characters such as protein phenotypes were expected to be more useful for this purpose because of the more direct relation between genes and proteins. Barley prolamins (hordeins) are associated with barley chromosome G (Islam *et al.*, 1975), which leads to sterility when added to wheat. The chromosomal locations of barley structural genes for alcohol dehydrogenase (ADH), a glutamic oxalacetic transaminase (GOT), an amino peptidase (AMP), an endopeptidase (EP) and some esterase variants (EST) are known, and these isozymes characterized four of the six addition lines (Hart *et al.*, 1981).

These isozymes proved to be very useful in the initial characterization of the disomic addition lines and they have been invaluable in more recent studies aimed at substituting particular barley chromosomes for specific wheat chromosomes.

The final characterization of the addition lines was achieved when the barley chromosomes present in the addition lines could be related to the standard numbering system adopted for barley chromosomes based on the numbering system used for the barley trisomics (Tsuchiya *et al.*, 1960). Islam (1980) applied a heterochromatic banding procedure (N-banding) to the somatic chromosomes of Betzes barley, Betzes trisomics and the addition lines. He found that each barley chromosome has a unique pattern and could be distinguished from the banded wheat chromosomes. By relating the banding pattern of the barley chromosome in the addition lines with the pattern of the extra chromosome present in each of the trisomic lines, the arbitrary capital letter designations (A,B,C,D,E,F,G) could be replaced with the standard numbering system used for barley chromosomes (4,7,6,1,2,3 and 5, respectively).

The regularity of chromosome pairing in the addition lines was similar to that of the wheat parent, except addition lines 1 and 4 showed more asynapsis at metaphase I. Under glasshouse conditions, all of the addition lines except line 1 had good fertility, and the only lines likely to need special care during their maintenance because of instability are lines 4 and 7 (Islam *et al.*, 1981).

In contrast to the more or less regular developmental behaviour of wheat plants containing these six barley chromosomes, those plants possessing barley chromosome 5 (including F_1 hybrids, BC_1 and some BC_2 progeny) exhibited many abnormalities during microsporogenesis (Islam and Shepherd, 1981a). These abnormalities included chromosome number mosaicism in PMCs, with hypoploid, hyperploid and highly polyploid cells, and some post-meiotic irregularities such as linear polyads and multipore pollen grains. The degree of abnormality present depended on the chromosome constitution of these plants; those with chromosome 5 alone were the most abnormal and were completely sterile. However, whenever barley chromosome 6 was present as well as chromosome 5, for example in a 44 chromosome double monosomic plant, the cytological behaviour was much less irregular and such plants were female fertile.

There is little doubt that barley chromosome 5, when added to wheat, results in modified spindle structure and function during premeiotic mitosis in archesporial cells and at meiosis in PMCs. The cytology closely parallels that induced in wheat by disruption of the spindle with colchicine at specific stages during the development of the anther (Dover, 1972; Dover and Riley, 1973). Furthermore, our observations are similar to some other findings of Dover (1973a,b) with wheat – *Ae. mutica* addition lines. He found that chromosome M of *Ae. mutica* when added to wheat induced homoeologous pairing at meiosis, chromosome mosaicism in PMCs and multipore pollen, and in the last two respects, at least, is similar to chromosome 5 of barley.

The question of whether chromosome 5 of barley is also associated with the irregular somatic chromosome constitution of most of the wheat x barley F_1 hybrids, remains unanswered. However, because the occurrence of abnormal chromosome constitutions is restricted to wheat x barley F_1 hybrids, and the abnormalities at microsporogenesis occur in any plants containing chromosome 5, irrespective of their source of cytoplasm, we believe these two abnormalities are probably due to different causes.

Utilization of the addition lines – The addition lines provide a new means of assigning barley genes to chromosomes, or chromosome arms, simply by determining which of the complete chromosome additions, or telosomic additions, expresses the barley character in question. This method will be useful for structural genes, but not for regulator or controlling genes, and of course only for those barley characters which are expressed in a wheat background.

Already, these lines have proved very useful for determining which barley chromosomes carry the structural genes for seed storage proteins (Islam *et al.*, 1981; Lawrence and Shepherd, 1981) and several isozymes (Hart *et al.*, 1981; Powling *et al.*, 1981) and the consolidated results are given in Table 8.3. The results obtained with the seed proteins of barley are of special interest because the prolamins (B2, B3, Figure 8.4) and glutelins (B1, Figure 8.4) are both controlled by chromosome 5. In the same study, the genes controlling glutelins were found to be on the long arm of chromosome 5, whereas the prolamin genes are on the short arm. Thus the chromosomal distribution of these genes in barley, resembles that in wheat where the group 1 chromosomes of each genome have genes controlling prolamins on their short arm and genes controlling glutelins on their long arm (Wrigley and Shepherd, 1973; Bietz *et al.*, 1975; Lawrence and Shepherd, 1980). These results indicate genetic similarities between chromosome 5 of barley and chromosome 1 of the wheat genomes, and their likely descent from a common ancestral chromosome. However, given this genetic similarity we must pose the question of what changes might have occurred in the products of these genes which allows the wheat glutelins in flour to aggregate into a visco-elastic gluten complex whereas barley flour seems to have no potential to develop visco-elastic properties.

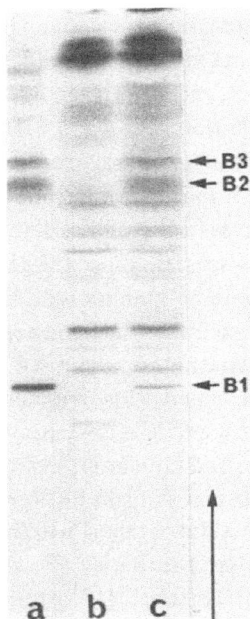

Fig. 8.4. SDS – PAGE patterns of total seed proteins from (a) Betzes barley, (b) Chinese Spring wheat, and (c) monosomic addition of Betzes chromosome 5 to Chinese Spring. (From Lawrence and Shepherd, 1981).

Table 8.3

Chromosomal location of structural genes for barley proteins

Barley chromosome	Protein character
1	EST-1, -2*; EP*
2	G-6-PDϕ
3	EST-3, -4*
4	ADH*, ACPHϕ, β amylaseϕ
5	Hordeins[†]; Glutelins[†]; PGI-1ϕ; MDHϕ
6	GOT-2*, AMP*
7	—

* Hart *et al.* (1981); [†] Lawrence and Shepherd (1981); ϕ Powling *et al.* (1981) — ACPH = acid phosphatase, PGI = phospho-glucose isomerase, MDH = malate dehydrogenase G-6-PD = glucose-6-phosphate dehydrogenase.

Homoeology of wheat and barley chromosomes - Evidence for genetic similarity between the chromosomes of wheat and barley has come from the addition lines. Hart *et al.* (1981) noted that the products of the barley and wheat genes controlling ADH isozymes, and also GOT-2 isozymes, were capable of forming active heterodimers in addition lines 4 and 6, and in the heptaploid, and concluded that the genes controlling the similar isozymes must be homoeologous. They thought it likely that the genes controlling the EP and AMP isozymes are homoeologous also, because of the similar developmental and tissue specificity of these isozymes in the two species. Since the gene locations for these isozymes in Betzes barley appeared to be syntenic with those in Chinese Spring wheat (Hart, 1979a), they suggested that the barley chromosomes 4,6 and 1 might be homoeologous with the wheat chromosomes 4,6 and 7 respectively.

Powling *et al.* (1981) compared the chromosomal location of genes controlling acid phosphatase and β amylase isozymes in barley and wheat (Hart and Langston, 1977; Joudrier and Cauderon, 1976) and reinforced the earlier suggestion that barley chromosome 4 might be homoeologous with chromosome 4 of the wheat genomes. Results obtained with malate dehydrogenase and phosphoglucose isomerase (PGI-1) in barley (Powling *et al.*, 1981) and wheat (Bergman and Williams, 1972; Hart, 1979b) lent support to the evidence from seed storage proteins (Lawrence and Shepherd, 1981) that chromosome 5 of barley shows at least partial homoeology with the group 1 chromosomes of wheat.

The barley chromosome 2 addition line resembles rye addition 2R (Riley and Chapman, 1958) and wheat tetrasomics 2A and 2D (Sears, 1954) in having narrow leaves, stems and spikes and this barley chromosome may also be homoeologous with wheat chromosomes of group 2.

We have produced lines having barley chromosome 6 substituted for wheat chromosomes 6A, 6B and 6D, and other lines having barley chromosome 4 substituted for wheat chromosome 4A. The genetic compensation of barley chromosome 6 for wheat chromosomes of group 6 has been excellent with respect to plant vigour and fertility. However, variable results have been obtained with barley chromosome 4 substituted for wheat chromosome 4A. The initial substitution was made using the addition line containing the heterobrachial form of barley chromosome 4, and the derived disomic substitution line showed excellent vigour and fertility, despite the general lack of vegetative vigour and complete male sterility of nullisomic 4A of wheat. This apparent ability of a barley chromosome to compensate well for chromosome 4A of wheat was rather surprising, because chromosome 4A has several unusual genetic and cytological features (reviewed by Driscoll, 1980), not least of which is the finding that its wheat homoeologues 4B and 4D when present in extra dose cannot compensate for the male sterility of nullisomic 4A (Sears, 1966). When we repeated this substitution test using barley chromosome 4 from a different source and having an apparently unmodified structure, the genetic compensation in the substitution line was much reduced. It is possible that the gene(s) for male fertility are located in the terminal segment of one arm of barley chromosome 4, as postulated for chromosome 4A of wheat (Barlow and Driscoll, 1980), and the chromosome used in the second substitution program may have lost this terminal segment.

Transfer of barley characters to wheat — The production of addition lines has made it possible, for the first time, to transfer barley characters to wheat using methods described by Sears (1956) and Riley *et al.* (1968) for transferring desirable characters from other alien species to wheat. The question of whether such transfers of barley genes into wheat will lead to improved wheat cultivars is difficult to answer. The first problem is to identify characters in barley which might confer some advantage to wheat and then it would be necessary to find whether the characters are expressed in a wheat background and whether they are controlled by genes on just one barley chromosome.

One character of interest is the reported tolerance of some barley cultivars to high levels of salinity (Epstein and Norlyn, 1977). Since wheat is susceptible to salinity, it would be advantageous if this barley character could be transferred and utilized in wheat. However, it is not known whether the tolerance to salinity will be expressed in a wheat background, nor is the genetic control of this character known. From our experience with morphological characters, it is likely that many agronomic and physiological characters of barley will not be expressed in the addition lines, especially if they are controlled by several genes each with small effect.

Another attempt at transferring a desirable barley character to wheat involves the possible transfer of resistance to barley yellow dwarf virus (Qualset, 1975, pers. comm.). Although it may not apply with this particular example, we believe that the widespread transfer of genes controlling disease resistance from barley to wheat, could be counter-productive since it would increase the genetic uniformity across crops and hence increase their genetic vulnerability.

Future prospects

Attempts have been made to hybridize wheat and barley for more than 80 years, but authentic hybrids have been available only since 1973. Progress has been rapid in the last seven years. Perhaps the most important question for the future is whether wheat:barley hybrids and their derivatives can contribute to the improvement of wheat as a crop plant. The success already obtained with such a wide cross as wheat x barley, involving members of two different sub-tribes according to Hutchinson (1934) or two widely different ecological groups of the Triticeae according to Sakamoto (1973), may lead to improvements in the wheat crop in an indirect way. Cytogeneticists may now be stimulated to attempt crosses between some other genera of the Triticeae which previously had not been crossed. Hybrids between these wild grasses and wheat may provide new sources of disease resistance for inclusion in wheat, without the likely disadvantages associated with the transfer of disease resistance genes from another cultivated crop.

The evidence of a close similarity between some wheat and barley genes (Hart *et al.*, 1981) and chromosomes is rather surprising considering the marked taxo-nomic differences between these crop plants (Hutchinson, 1934), their separate domestication 10 000 years ago (Harlan, 1968; Riley, 1975) and their separ-ation from a common ancestor possibly as long ago as in the Miocene-Pliocene epochs of the Tertiary period (Sakamoto, 1973). Much valuable information on the evolutionary relationship between wheat and barley will come from further determinations of the genetic equivalence of their chromosomes.

We are optimistic that a fertile amphiploid will be produced from wheat x barley hybrids in the future, especially since one has already been obtained with wheat x *H. chilense* (Martin and Chapman, 1977). Success may depend on find-ing a variant of chromosome 5 of barley which does not cause spindle abnormal-ities when added to wheat, and we are now searching for it among wild and cult-ivated barley species. If this search is successful, we might be able to produce not only the amphiploid but also the disomic addition line carrying barley chromosome 5, which is required to complete the set of wheat-barley addition lines. Another research priority is to find the origin and cause of the variation in somatic chromosome constitution which occurs in wheat x barley hybrids. More information is needed on the cytological basis of the irregular chromosome distribution suspected to occur during early divisions of the zygote. Also we need to know whether environmental factors such as changes in day length, or genetic factors, such as the presence of one or more barley chromosomes in the ovule parent, might overcome this problem. It is important to resolve this question because currently it is the main factor restricting the production of new wheat x barley hybrid combinations.

References

Ahokas, H. (1970) Some artificial intergeneric hybrids in the Triticeae. *Ann. Bot. Fenn.* 7, 182–192.

Barclay, I.R. (1975) High frequencies of haploid production in wheat (*Triticum aestivum*) by chromosome elimination. *Nature (Lond.)* **256**, 410–411.

Barlow, K.K. and Driscoll, C.J. (1980) Linkage studies involving two chromosomal male-sterility mutants in hexaploid wheat. *Genetics* (in press).

Bates, L.S., Campos, V.A., Rodriguez, R.R. and Anderson, R.G. (1974) Progress toward novel cereal grains. *Cereal Sci. Today* **19**, 283–286.

Bates, L.S., Mujeeb, K.A. and Waters, R.F. (1976) Wheat x barley hybrids problems and potentials. *Cereal Res. Commun.* **4**, 377–386.

Bergman, J.W. and Williams, N.D. (1972) Isozyme variants of esterase and malate dehydrogenase among wheat aneuploids. *Agronomy Abstracts* p. 23.

Bietz, J.A., Shepherd, K.W. and Wall, J.S. (1975) Single-kernel analysis of glutenin: use in wheat genetics and breeding. *Cereal Chem.* **52**, 513–532.

Chapman, V. and Miller, T.E. (1978) Barley wheat hybrids. Ann. Report Plant Breed. Inst. Camb. 1978. pp. 123–124.

Dover, G.A. (1972) The organization and polarity of pollen mother cells of *Triticum aestivum*. *J. Cell Sci.* **11**, 699–711.

Dover, G.A. (1973a) On pores and pairing. *Chromosomes Today* **4**, 197–204.

Dover, G.A. (1973b) The genetics and interactions of 'A' and 'B' chromosomes controlling meiotic chromosome pairing in the Triticinae. Proc. 4th Int. Wheat Genetics Symp. Columbia, Mo. pp. 653–666.

Dover, G.A. and Riley, R. (1973) The effect of spindle inhibitors applied before meiosis on meiotic chromosome pairing. *J. Cell Sci.* **12**, 143–161.

Driscoll, C.J. (1980) Perspectives in chromosomal manipulation. In 'The Manipulation of Genetic Systems in Plant Breeding'. Royal Society Discussion Meeting London, 1980.

Epstein, E. and Norlyn, J.D. (1977) Seawater-based crop production: A feasibility study. *Science* **197**, 249–251.

Farrer, W. (1904) Some notes on the Wheat "Bobs"; its peculiarities, economic value and origin. *Agric. Gaz. N.S.W.* **15**, 849–854.

Fedak, G. (1977) Increased homoeologous chromosome pairing in *Hordeum vulgare* x *Triticum aestivum* hybrids. *Nature (Lond.)* **266**, 529–530.

Fedak, G. (1980) Production, morphology and meiosis of reciprocal barley-wheat hybrids. *Can. J. Genet. Cytol.* **22**, 117–123.

Gordon, G.S. and Raw, A.R. (1932) Wheat-barley matings. *Victorian Department of Agriculture Journal (Australia)* **30**, 138–144.

Harlan, J.R. (1968) On the origin of barley. In 'Barley: Origin, Botany, Culture, Winter Hardiness, Genetics, Utilization, Pests'. U.S.D.A. Agric. Handbook, No. 338, 1968 edn. pp. 9–31.

Hart, G.E. (1979a) Genetical and chromosomal relationships among the wheats and their relatives. *Stadler Genet. Symp.* **11**, 9–29.

Hart, G.E. (1979b) Evidence for a triplicate set of glucose phosphate isomerase structural genes in hexaploid wheat. *Biochem. Genet.* **17**, 585–598.

Hart, G.E. and Langston, P.J. (1977) Chromosomal location and evolution of isozyme structural genes in hexaploid wheat. *Heredity* **39**, 263–277.

Hart, G.E., Islam, A.K.M.R. and Shepherd, K.W. (1981) Use of isozymes as chromosome markers in the isolation and characterization of wheat-barley chromosome addition lines. *Genet. Res.* (in press).

Hutchinson, J. (1934) 'The Families of Flowering Plants. II. Monocotyledons.' McMillan and Co. Ltd., London. pp. 207–208.

Islam, A.K.M.R. (1980) Identification of wheat-barley addition lines with N-banding of chromosomes. *Chromosoma (Berl.)* **76**, 365–373.3.

Islam, A.K.M.R. and Shepherd, K.W. (1980) Meiotic restitution in wheat-barley hybrids. *Chromosoma (Berl.)* **79**, 363–372.

Islam, A.K.M.R. and Shepherd, K.W. (1981a) Cytological abnormalities in wheat:barley hybrids and their derivatives (submitted for publication).

Islam, A.K.M.R. and Shepherd, K.W. (1981b) Production of disomic wheat-barley chromosome addition lines using *Hordeum bulbosum* crosses. *Genet. Res.* (in press).

Islam, A.K.M.R., Shepherd, K.W. and Sparrow, D.H.B. (1975) Addition of individual barley chromosomes to wheat. Proc. 3rd Int. Barley Genetics Symp. Garching, W. Germany. pp. 260–270.

Islam, A.K.M.R., Shepherd, K.W. and Sparrow, D.H.B. (1978) Production and characterization of wheat-barley addition lines. Proc. 5th Int. Wheat Genetics Symp. New Delhi, 1978. pp. 365–371.

Islam, A.K.M.R., Shepherd, K.W. and Sparrow, D.H.B. (1981) Isolation and characterization of euplasmic wheat-barley chromosome addition lines. *Heredity* (in press).

Joudrier, M.P. and Cauderon, Y. (1976) Localisation chromosomique de genes contrôlant la synthèse de certains constituants β amylasique du grain de Blé tendre. *C.R. Hebd. Seances Acad. Sci. Ser. D Sci. Nat.* **282**, 115–118.

Kimber, G. and Sallee, P.J. (1976) A hybrid between *Triticum timopheevi* and *Hordeum bogdanii. Cereal Res. Commun.* **4**, 33–37.

Kruse, A. (1973) *Hordeum* x *Triticum* hybrids. *Hereditas* **73**, 157–161.

Kruse, A. (1976) Reciprocal hybrids between the genera *Hordeum, Secale* and *Triticum. Hereditas* **84**, 244.

Lawrence, G.J. and Shepherd, K.W. (1980) Variation in glutenin protein sub-units of wheat. *Aust. J. Biol. Sci.* **33**, 221–233.

Lawrence, G.J. and Shepherd, K.W. (1981) Chromosomal location of genes controlling seed proteins in species related to wheat. *Theor. Appl. Genet.* (in press).

Macindoe, S.L. (1939) William Farrer's contribution to our knowledge of inheritance. *J. Aust. Inst. Agric. Sci.*. 5, 208–212.

Macindoe, S.L. and Brown, C.W. (1968) Wheat breeding and varieties in Australia. Science Bull. No. 76 N.S.W. Dept. of Agriculture 3rd edn. p. 46.

Martin, A. and Chapman, V. (1977) A hybrid between *Hordeum chilense* and *Triticum aestivum. Cereal Res. Commun.* 5, 365–368.

O'Mara, J.G. (1940) Cytogenetic studies on Triticale. I. A method for determining the effects of individual *Secale* chromosomes on *Triticum. Genetics* 25, 401–408.

Powling, A., Islam, A.K.M.R. and Shepherd, K.W. (1981) Isozymes in wheat-barley hybrid derivative lines. *Biochem. Genet.* (in press).

Pridham, J.T. (1914) New varieties of wheat. *Agric. Gaz. N.S.W.* 25, 230–233.

Pye, H. (1921) Report by Hugh Pye, Cerealist, Dookie Agricultural College Experiment Station. Council of Agricultural Education. Government Printer, Melbourne.

Qualset, C.O. (1975) Sampling germplasm in a center of diversity : an example of disease resistance in Ethiopian barley. In 'Crop Genetic Resources for Today and Tomorrow'. (Eds. O.H. Frankel and J.G. Hawkes.) Cambridge University Press. pp. 81–96.

Riley, R. (1975) Origins of wheat. In 'Bread'. Proc. of the Rank Prize Funds Int. Symp. 1974. Appl. Sci. Pub. Ltd. pp. 27–45.

Riley, R. and Chapman, V. (1958) The production and phenotypes of wheat-rye chromosome addition lines. *Heredity* 12, 301–315.

Riley, R., Chapman, V. and Johnson, R. (1968) The incorporation of alien disease resistance in wheat by genetic interference with the regulation of meiotic chromosome synapsis. *Genet. Res.* 12, 199–219.

Sakamoto, S. (1973) Patterns of phylogenetic differentiation in the tribe Triticeae. *Seiken Ziho* 24, 11–31.

Sears, E.R. (1954) The aneuploids of common wheat. *Mo. Agric. Exp. Stn. Res. Bull.* 572. 58 pp.

Sears, E.R. (1956) The transfer of leaf-rust resistance from *Ae. umbellulata* to wheat. *Brookhaven Symp. Biol.* 9, 1–22.

Sears, E.R. (1966) Nullisomic-tetrasomic combinations in hexaploid wheat. In 'Chromosome Manipulations and Plant Genetics'. (Eds. R. Riley and K.R. Lewis.) A supplement to *Heredity* 20, 29–45.

Sims, H.J. (1980) Wheat breeding in Victoria — some historical notes. Suppl. to Treatise of Third Assembly of the Wheat Breeding Society of Australia Inc. Victorian Crops Research Inst. Longerenong Agric. College, Aug. 1980. 33 pp.

Thomas, J.B., Mujeeb, K.A., Rodriguez, R. and Bates, L.S. (1977) Barley x wheat hybrids. *Cereal Res. Commun.* **5**, 181—188.

Tsuchiya, T., Hayashi, J. and Takahashi, R. (1960) Genetic studies in trisomic barley. II. Further studies on the relationship between trisomics and the genetic linkage groups. *Jpn. J. Genet.* **35**, 153—160.

Waterhouse, W.L. (1930) Australian rust studies. III. Initial results of breeding for rust resistance. *Proc. Linn. Soc. N.S.W.* **55**, 596—636.

Wrigley, C.W. and Shepherd, K.W. (1973) Electrofocusing of grain proteins from wheat genotypes. *Ann. N.Y. Acad. Sci.* **209**, 154—162.

9   WHEAT AND ITS RUST PARASITES IN AUSTRALIA

I.A. Watson

Introduction

Wheat in Australia may be attacked by three different species of rust fungi. Two of these are well known and have probably been here since our earliest settlement. The third species *Puccinia striiformis* West., causing stripe rust, was detected for the first time in 1979 (O'Brien *et al.*, 1980). At present the immediate threat from rust parasites of wheat in the traditional rust-liable areas of north eastern Australia is minimal and no significant losses have been reported there for over 30 years. This situation has resulted from a modest but effective research and breeding programme that has utilized the latest genetic information on variation in the host and the parasites.

In some of the southern areas little breeding work to control rust has been carried out because the losses over the last 60 years have been sporadic. These are now in fact the real danger spots because many of the predominant cultivars are susceptible and little farm hygiene is practised. This situation was highlighted in the recent serious rust epiphytotic of 1973-74 when crops in the southern states were severely damaged by stem rust causing a substantial loss of yield. It is anticipated that with the implementation of the National Rust Control Programme the probability of a similar happening in the future will be reduced.

It is still too early to estimate the potential damage that could result following the arrival of *P. striiformis*. Stripe rust continues to be a problem in Europe, India, in Central and South America and in other parts of the world. The climatic conditions in these regions are different from those in the Australian wheat belt and the experiences in other countries are of little help in predicting how the organism will colonize here. Should the disease become serious, then one can confidently expect that it too can be contained by a planned breeding strategy. Such a strategy has been evolved and found to work with other rusts and I will trace the highlights observed during that evolution using stem rust as an example.

Early studies

It is still not known how and when the two wheat rust organisms, *Puccinia graminis* Pers.f.sp. *tritici* Eriks. & Henn. (stem rust) and *Puccinia recondita* Rob.ex Desm.f.sp. *tritici* Eriks & Henn. (leaf rust) arrived in Australia. They probably preceded the first migrant settlers and survived on endemic graminaceous hosts. The first satisfactory report that the blights and mildews referred to in the popular press were probably rusts concerned a crop at Brush Farm, Eastwood, N.S.W. in 1803 (Waterhouse, 1936). No studies of rust were undertaken in Australia for many years despite the spectacular advances that were being made in Europe during the nineteenth century. There were no mycologists of any consequence in the country at that time and, although Farrer could recognize the different symptoms caused by *P. graminis tritici* and *P. recondita*, the reports of the five 'Rust in Wheat' conferences held prior to 1900 showed that little was known of the diseases or of the fungi that caused them.

Scientific experimentation with wheat rust organisms was started by Waterhouse about 1920, so we are now celebrating 60 years of uninterrupted wheat rust research in Australia. Throughout the period much of the work has related to variability within the organism, its origin and importance both in the survival of strains and in the stability of the reaction of resistant cultivars. The political homogeneity and relative isolation of Australia compared with other continents where wheat is grown have provided unique opportunities for the study of vast populations of these organisms, subjected to a minimum of contamination from spores originating elsewhere.

Australia and New Zealand were found by Waterhouse to comprise a single geographical area with wind-borne uredospores flowing freely across the continent and the Tasman Sea. Appearance of new rusts in Australia, not only of *P. graminis* and *P. recondita* but others as well have been paralleled by their occurrence some time later in New Zealand.

There is evidence during the survey period for three separate and independent occasions when inoculum, presumably as wind-borne uredospores, arrived from other geographical areas of the world. This served to provide the necessary materials to study what happens when new genes for virulence or avirulence are introduced into already well established populations.

The rust organisms comprise many parasitic strains alike morphologically but distinguished by their combinations of genes for pathogenicity. In the original studies of *P. graminis tritici* Waterhouse found 6 standard races; they were apparently typical of the area since they had not been reported elsewhere at the time. They differed in frequency, appeared to be little influenced by the pattern of the wheat cultivars grown, and for the first 5 years of these studies no dramatic change in the population structure occurred. Little was known about genes for specific resistance and consequently it was almost impossible to recognise mutant strains as they are known today. In the first attempts at breeding for stem rust resistance, Waterhouse combined different genes in Canberra and Thew and obtained the cultivar Euston resistant to all six races. It was never grown commercially.

Results from this first breeding effort were upset because sometime in 1925 Euston became susceptible to a new type of *P. graminis tritici* whose arrival in Western Australia was to have significant effects on the local rust population over the next 15 years. Moreover, it also influenced all future breeding work. It was so different in pathogenicity and in reproductive potential (McIntosh and Watson, 1981) from any of the earlier races that it was unlikely to have evolved from them. *Berberis vulgaris* L., the alternate host, had never been found rusted in the west and a mutation of such complexity from the original population was most unlikely. The latter appeared to be fixed for a recessive gene for virulence on plants with *Sr21* in Einkorn but the new rust carried the dominant allelic gene for avirulence. Waterhouse believed it was a new arrival and he was probably right. Owing to its marked competitive ability it overran the other six races and by 1940 it was almost impossible to isolate any strain other than the new one now designated 126-5,6,7,11[*].

Further attempts at breeding

Prior to 1926 little valuable stem rust-resistant material was available for use by breeders. There was no single line of *T. aestivum* L. that gave protection against all strains. However, work had been proceeding in the United States with this species, and Hope and Webster became available. Iumillo (*T. turgidum* L.) had also been used in the derivation of Marquillo. Breeding work with the first two was started immediately and since Hope had adult plant resistance in the field to many strains, but seedlings may be susceptible, its use started a controversy which continues today. Some plant pathologists and breeders favour this type of parent which shows adult plant resistance in preference to one whose seedlings and adult plants are both resistant when the same strain of rust is involved. This preference stems from a belief that seedling resistance (mainly hypersensitivity) readily succumbs to the depredations of mutants and recombinants which acquire new virulence.

[*]The first number in the designation is the standard race number and the following numbers 1,2,3,4,5,6,7,11 represent virulence respectively on plants having genes *Sr6, Sr11, Sr9b, SrTt₁, Sr17, Sr8, Sr15* and *Sr* Barleta Benvenuto.

The use of Hope resulted in two new resistant cultivars being released around 1940, Hofed and Warigo. Both have remained resistant to this time. There was greater opportunity to exploit the resistance due to hypersensitivity because about 1930 the N.S.W. Department of Agriculture introduced an extensive collection of wheats from Kenya Colony, East Africa. These wheats had gene-specific resistance because the genes involved in it could be isolated and recognized by any race or group of races in which the corresponding gene for avirulence occurred. It was not known in 1930 but in these early Kenya wheats and in another introduction, Gaza (*T. turgidum*), breeders were using three genes for specific resistance *Sr6, Sr9b* and *Sr11*.

Because *Sr6* controls near-immunity in the field when the appropriate strains are present, Macindoe concentrated on its use, and he also used *Sr11*. Waterhouse on the other hand found that seedling resistance due to *Sr6* was not uniformly effective against all strains, and he concentrated on *Sr11* from the Kenya wheats and made minor use of *Sr9b*. He also used *Sr11* from Gaza because in addition it has *Lr23* for resistance to *P. recondita*. It is now well known that Eureka (*Sr6*) evolved from the first programme and was released in 1938, while Gabo (*Sr11 Lr23*) (1945) and a number of other wheats came from the second.

Eureka was selected against the single strain 126-5,6,7,11. It was highly successful, and well received by farmers so that in 1940 and 1941 it appeared to have solved the problem of stem rust. The reasons for its failure to do so have been well documented (Watson and Luig, 1963). While this was a disappointment, other wheats were multiplied as a replacement and most had *Sr11*; Gabo, Kendee, Yalta and Charter were released to N.S.W. about this time and Macindoe released the Iumillo derivative, Celebration. These wheats kept rust under control until 1948-49 when all but Celebration became susceptible because of the appearance of two new strains. With hindsight, these events can now be clearly explained - and they underline the importance of mutation as a basis for variation in rust fungi.

As Figure 9.1 suggests, the 1925 strain had two avirulence genes *P6* and *P11*. Each mutated independently to give other strains so that by 1948 four basic strains were present in the field. There were some slight variations noted when these strains were tested on the differential series. Acme and Kubanka often showed a lower infection type resulting in a change in race designation from 126 to 222.

These simple mutational processes observed at this time seriously called into question the breeding techniques being employed. It was asked if breeding work using single gene resistances was satisfactory. Clearly more studies were needed on the organisms themselves.

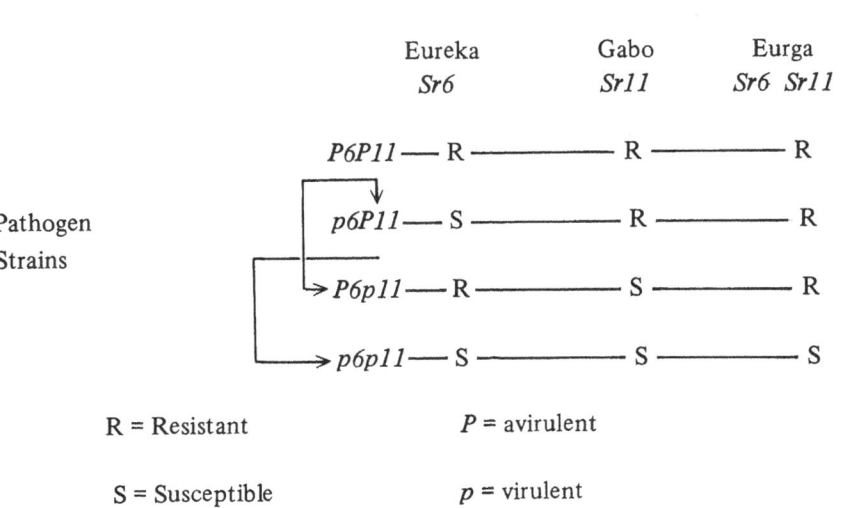

Fig. 9.1. The probable origin of strains of *P. graminis* virulent on plants with *Sr6* and *Sr11*.

The processses involved in pathogenicity changes

(a) Sexual reproduction

In order to formulate breeding procedures likely to succeed in the light of these sudden changes in pathogenicity it was necessary to examine the basis for such variation. Both sexual and asexual processes could be involved. For a long time it was known that *Berberis vulgaris* and *Thalictrum flavum* L. were the alternate hosts for *P. graminis* and *P. recondita*, respectively. The extent of the participation of these species in pathogenicity changes was, however, unknown before Waterhouse carried out his important studies in the late 1920's. He showed that in those cases where clonal material is heterozygous for avirulence the passage of the respective organisms through the alternate host results in the recovery of individual cultures with a host range mostly wider than that of the parent culture. However, he did not realise that he was observing the segregation of homozygous recessive genes for virulence.

Experiments similar to those of Waterhouse had also been made abroad but with cultures homozygous for pathogenicity on wheats having certain genes. Hence no segregates appeared among the selfed progeny with a host range wider than that of the parent. The Australian research helped to justify a continuation of the massive barberry eradication programme that had been mounted in the

United States. The origin of new virulence gene combinations resulting from
sexual reproduction had some practical significance in the case of *P. graminis
tritici*, but it was only of academic interest in *P. recondita* because *Thalictrum*
infected by that organism in the field had not been found in Australia or the
United States. Hence the widespread variation in both *P. graminis tritici* and
P. recondita could not be attributed to sexual reproduction.

(b) Mutation

The first indication that asexual variation could occur in Australian rusts was
in *P. recondita* in the early 1920's. Waterhouse found two types of this organism, one was virulent on Thew (*Lr20*) the other avirulent, otherwise they were
identical. These were later designated races 26 and 95 respectively (Waterhouse,
1952). When working with a pure culture of race 95 it was invariably found that
isolated virulent pustules occurred on Thew and these were always of race 26.
P20 is very unstable and the variants were single gene mutations from avirulence
to virulence.

In 1945 Gabo (*Lr23*) was widely grown and in that year isolated pustules
were found on the leaves for the first time. Prior to this, races 26 and 95 were
the only ones known in the field but Waterhouse (1952) described two different
standard races, 135 and 138 from Gabo. These presumably arose from the
parents 26 and 95 but were not single gene mutations since they differed from
them at more than one locus. We have found such mutants to be very uncommon.

Spontaneous mutations from avirulence to virulence in certain genes can be
readily selected in the glasshouse (Watson, 1957a). The fungal genes differ
significantly in the rates at which they mutate but the ranking under natural and
artificial conditions is about the same (Flor, 1958). Luig (1978) has enlarged on
this proposition in so far as it applies to *P. graminis tritici* and we have jointly
worked for many years on the occurrence of variants in the field and glasshouse.
A general summary of these findings suggests that the fungal genes may be
classified as follows:

(i) Genes which mutate readily under natural conditions and following treat-
     ment with chemical mutagens such as E.M.S. Representative of this
     group are *P5, P9e* and *P21*.

(ii) Genes which mutate under natural conditions but which appear not to be
     influenced by mutagens such as E.M.S. The gene PTt_1 probably falls
     into this category. When $SrTt_1$ was released singly in Mengavi in 1958 it
     gave spectacular resistance to all field strains known at the time, mainly
     21-2, 21-5 and 34-2. After about three years of commercial cultivation

strains 21-4,5 and 34-2,4, both virulent on Mengavi, caused the discontinuation of further plantings. In view of this it is surprising that Luig (1978) had difficulty in producing variants on plants with $SrTt_1$ using the post-1954 isolates. The pre-1954 rust, 126-5,6,7,11, readily produced mutants on such plants following the same treatment and mutants also appeared in the field (Watson, 1955).

(iii) Genes which have a low natural mutation rate and which are not influenced by mutagens such as E.M.S. Genes *P13, P24* and *P27* are suggested for this group.

(iv) Genes which do not mutate or do so at a very low rate. *P26* fits this category and hence *Sr26* from *Agropyron elongatum* (host) Beauv. continues to be an effective source of resistance.

As will be shown later induced mutations have been important in the breeding programme in that they have facilitated selection against strains with a wider host range. These are the anticipated ones which have not yet appeared in the field.

(c) Progressive changes in virulence

Reports of mutations in pathogenicity mostly concern changes in cultures avirulent on plants having a specific gene to those having full virulence. But careful examination of large numbers of cultures from field surveys shows that some suspected of being fully virulent actually represent gradations in pathogenicity between avirulence and virulence. It can be queried if conversion from avirulence to virulence proceeds in one step. In initial experiments conducted by Watson and Luig (1968) it was found that by selecting for increased sporulation in cultures avirulent on plants with *Sr11*, derivatives could be isolated having various grades of infection type between the original avirulence level and the fully virulent type. Since more than three levels of phenotypic expression were observed, the changes were not due to a state of heterozygosity in the dikaryon and were described by the authors as progressive changes in virulence.

McIntosh (1977) reported that variation of this type now appears to be typical of many *Sr* loci in wheat and in the case of *Sr15*, five different levels of interaction between host and parasite can be recognized. He exposed plants having this gene to E.M.S. and selected for infection types distinguishable from those shown by control plants when inoculated with a culture giving the lowest of the five levels of interaction. Genetic analysis of the mutant derivatives with intermediate infection types demonstrated that the induced changes had occurred within the *Sr15* gene (McIntosh and Watson, 1981). To some extent these studies complement those of Watson and Luig but unfortunately it has not been possible to determine the genetic basis of the progressive virulence increases in the pathogen.

(d) Somatic hybridization

By 1950 variants whose origin could be traced to single gene mutations had been found in many parts of the world, but other changes were occurring from time to time that were not so easily explained. Research into other processes was stimulated following the work of Nelson *et al.* (1955) in the United States. They showed that pure cultures could occasionally give rise to others with wider virulence and suggested that variation in the normal dikaryotic arrangement of nuclei within the cell was involved. Australian work on this subject stemmed from a finding made by the author when at the University of Minnesota which indicated that if certain cultures of *P. graminis tritici*, but by no means all, were mixed together on congenial seedlings they underwent a change and variants arose (Watson, 1957b). The initial experiment involved the careful selection of two United States cultures, one yellow (NR-2) but apparently homozygous for many recessive alleles for virulence, the other red (111) and having a number of contrasting heterozygous genes. Both cultures had been collected from barberry and red (111) probably arose as a sexual hybrid between *P. graminis tritici* and *P. graminis secalis*.

From the initial experiment four new red cultures were recovered. They were so marked genetically that laboratory contamination could be ruled out as a possibility in their origin and the populations were too small for mutation to have been a factor. Using these same two parental cultures and making further mixtures, at least 20 new types have been derived since the work began 25 years ago. Moreover, the same technique has been used to show the role of somatic hybridization in the field (Watson and Luig, 1958).

In the same way it has been possible to demonstrate the asexual origin of variants arising from the mixing of two *formae speciales*, viz. *P. graminis tritici* and *P. graminis secalis*. Cultures of these were mixed and placed on a number of wheat genotypes and the derived hybrids were unlike any rust cultures with which we had previously worked (Watson and Luig, 1959).

The exact nature of the changes within the cells that enable many new strains to arise when two cultures are mixed is not known. However the diversity in the resultant progeny suggests that parasexualism, possibly of a type found in *Aspergillus*, may be involved. This would necessitate a diploid phase in the clonally propagating material. So far this has not been found but Williams (1975) reported a monokaryotic strain of *P. graminis tritici* established by Maclean *et al.* (1971) to be diploid. This strain, S.U. culture 71868 first observed in axenic culture, grows normally on wheat seedlings.

Williams compared the size of the nuclei in both the monokaryon and parental dikaryon after growing both cultures on wheat seedlings. He found that the nucleus of the former had a diameter one third greater than that of the nucleus of the latter, and it had about twice the volume. This relationship in volume corresponds approximately to that found when the diploid fusion nucleus of a teleutospore is compared with a haploid nucleus in the uredospore of the parent culture 334. The diameter of the fusion nucleus was also greater than that of the haploid. As further evidence to suggest the diploid nature of the mono-karyon nucleus, Williams and Mendgen (1975) compared the DNA content of the nuclei of the above two cultures using Feulgen fluorescence photometry and concluded that in the monokaryon it was double that in the single haploid nucleus of the parental dikaryon.

While this evidence for diploidy in some uredospores is clear, there is no information to show that the diploid culture 71868 readily undergoes a haploid-ization process, at least while growing on the host. In extensive experiments with large populations of uredospores and involving many thousands of infected wheat seedlings, the author found no tendency for the diploid to dissociate into dikaryons with a diversity of combinations of genes for virulence. Also in large scale paired mixture experiments using culture 71868 and several normal dikary-otic strains, no somatic hybrids were obtained. The only variation was the appearance of monokaryon and presumably diploid mutants with virulence on plants having one or other of the recognized *Sr* genes used for testing pur-poses.

From the work so far it must be concluded that under local conditions somatic hybridization of a type resembling parasexualism may take place when certain strains occur together on a congenial host, but survival of derivatives in the field is a rare event.

The role of exotic combinations of virulence genes

It is clear from the above that processes of unequal importance contribute to variation in virulence in the rust populations of Australia. By far the most important is mutation. However from time to time there are infusions of new genes and new gene combinations into the local rust flora. These are believed to originate in overseas inoculum which is presumed to be aerially borne either in wind currents or in wheel bays of large aircraft arriving from overseas. Whatever the place of origin and the mode of entry, new rusts have quickly adjusted to the new environment and Luig (1977) has outlined some of the factors involved in their establishment.

Figure 9.2 presents a graphical representation of some of the events that have taken place in the Australian stem rust flora over the past 60 years. Many of the standard races from those we have isolated are omitted. Waterhouse showed that the original six comprised two groups each of three. They were replaced

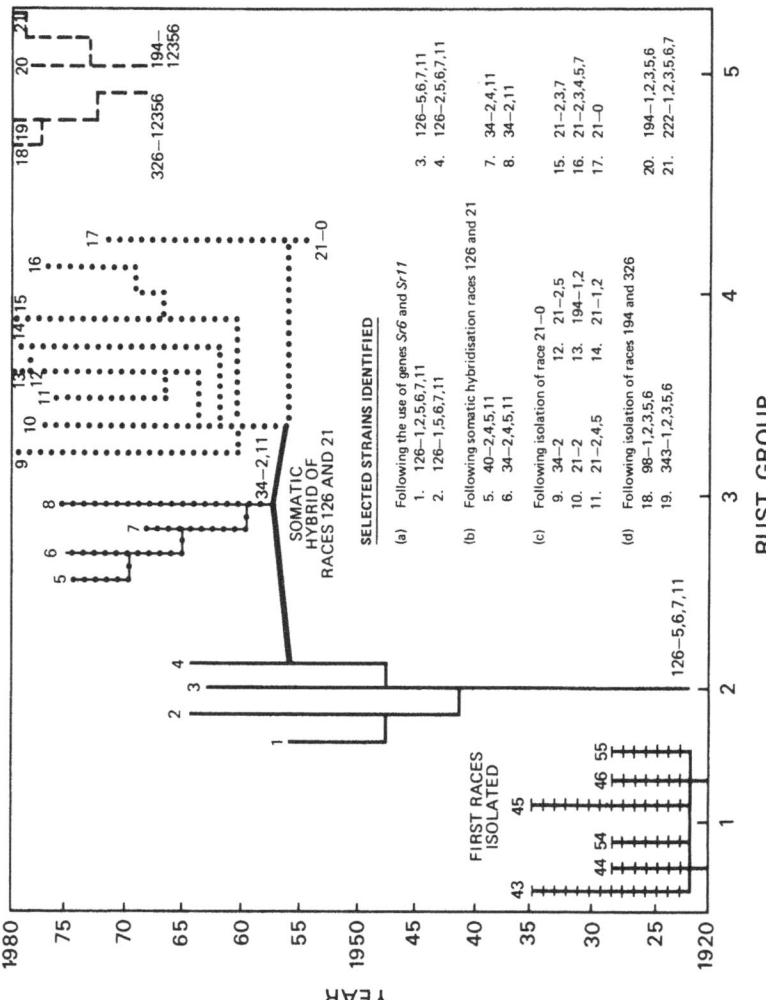

Fig. 9.2. A diagrammatic representation of the possible origin and evolution of selected strains of *P. graminis tritici* in Australia from 1920–1980.

over the next 15 years from 1925 by what Waterhouse called race 34 but which has been reclassified as 126-5,6,7,11 and this elimination resulted in essentially only one strain for the breeders to combat in the 1930's. Mention has already been made of the release of Eureka, the appearance of a mutant virulent on it, subsequent release of Gabo and the appearance of still further mutants on it in 1948-49. Few resistant cultivars remained, so these events put the Australian crop in a very vulnerable position. Warigo and Celebration with good resistance were available but their yielding abilities were below requirements. Glenwari lacked the grain quality that was considered essential.

As new breeding programmes were launched to broaden the genetic base for resistance against the predominant rusts (the 126-222 group), an event in 1954 changed the whole scene. In that year strain 21-0, totally unrelated to any rust previously found in the country, was isolated in southern N.S.W. and in Victoria. In competitive ability it completely outclassed its predecessors, including the up-till-then aggressive 126-5,6,7,11, and within eight years had eliminated them.

Watson and Luig (1966) have already listed a number of new characters that made this strain so different from any other previously studied in Australia. Among these were virulences on plants with *Sr9g* and *Sr12* and avirulence on plants with *Sr15, Sr17* and on Barleta Benvenuto (*SrBB*). Quite important too was the virulence on Arnautka, Mindum and Spelmar at low temperatures (16°C). While all these specific characters have been vital in our studies to monitor the Australian rust population since 1954, the over-riding influence that strain 21-0 and its derivatives were to have over the next 25 years was due to its competitive ability.

Strain 21-0 was avirulent on Eureka and Gabo, but a new derivative 21-2 soon appeared to allow colonization in the areas where cultivars with *Sr11* were grown. Eureka had declined in popularity and strains virulent on it largely disappeared. Hence it was again resistant. As an experiment in 1957 Eureka was recommended for cultivation for a second time, the first being in 1939. Farmers cautiously accepted it on a small scale but new mutations now in the background gentoype of standard race 21 arose to attack it (Watson and Luig, 1963). This sequence of events refuted any argument that favoured the recycling of single genes for resistance as a means of controlling stem rust. Several cultivars with a narrow base for resistance, e.g. Glenwari (*Sr17*), Mengavi (*SrTt1*) and Gamenya (*Sr9b*), all became susceptible about this time to further variants of the race 21 group. Evolution of new strains was rapid and over 50 mutational variants within this group were found during the 1960's (Luig and Watson, 1970).

The dramatic events which started in 1925 and in 1954 were paralleled in 1968/69 by the arrival of two more putative migrant races, 194 and 326. Already there is evidence that these will cause the third major shift in the Australian population of *P. graminis tritici*. Their isolation was assisted by the fact that at this time cultivars with *Sr6* were highly resistant in the field on account of a decrease in the frequency of strains with the gene *p6*. Large pustules on a line with *Sr6* at Clinton, S. Australia and on Bowie at Tichborne N.S.W., proved to be of standard races 326 and 194, respectively. They were well marked genetically and both showed very low infection types on plants with either *Sr9g* or *Sr15*. In addition race 326 was avirulent on plants with *Sr7b*. It has been easy to watch the development of both types over the last 10 years. While there is some doubt concerning the origin of races 126 and 21, all the evidence now suggests that they originated overseas. In the case of 194 and 326 the data available are more conclusive that the spores arrived from an African country. The first clue for this came from the International Virulence Gene Survey (Luig, unpublished) in which data on the rusts from Mozambique, available from Portugal, showed the presence of standard races 194 and 222 in this area. The survey made it clear that the rusts resembled those in Australia. Subsequent detailed work at Sydney University by de Sousa (1975) showed that four rusts in Australia 194-1,2,3,5,6, 194-1,2,3,5,6,7, 343-1,2,3,5,6 and 222-1,2,3,5,6 were present in Angola and some if not all were in Rhodesia. Race 326-1,2,3,5,6 was not found but 343-1,2,3,5,6 is a mutant from it with virulence on plants with *Sr5*.

The comparative studies were based on the characters of spore size, growth in axenic culture and, of course, on pathogenicity. All results showed that the rusts from the two continents were inseparable. The avirulence on Marquis (*Sr7b*) found in the Australian rusts 326 and 343 was confirmed in the Rhodesian rust in Salisbury, Rhodesia by the author in 1975. It is interesting that the alleles for both virulence and avirulence on plants with *SrBB* are present in Africa and each may have reached Australia from there.

While races 194 and 326 were the first to be recognized in this new cycle 10 years ago, some diversity has since developed in each of them. Variants resulting from the marked instability at the *P5* locus have arisen to give 222-1,2,3,5,6 (from 194) and 343-1,2,3,5,6 (from 326) as the first step in the adjustment to the Australian environment. It must be made clear that the isolation of race 222 as a derivative from 194 does not mean a return to the 126-222 race group. The latter, as shown in Table 9.1, carries a temperature-sensitive gene controlling pathogenicity on Arnautka, Mindum and Spelmar. Components of it are avirul-ent on these genotypes at low temperatures (16°C) but become virulent as the temperature rises. Race 222 derived from 194 by contrast is virulent on these genotypes through a wide range of temperatures. Moreover, rusts of the 126-222 group were last isolated in 1962.

Table 9.1

Reaction of Arnautka, Mindum and Spelmar to rust races at different temperatures.

| | Races and Period | | | |
| | 1942–62 | | 1968–80 | |
Temperature	126	222	194	222
16°C	;	;	4	4
19°C	X	X	4	4
23°C	4	4	4	4

; = Resistant   X = Mesothetic   4 = Susceptible

While race 194 has so far undergone one change, 326 first mutated to produce 343 a race first collected at Kerang, Victoria, in 1973. Race 98-1,2,3,5,6 is from the second change in which 343 has mutated to virulence on plants with *Sr9g*. The significance of this is not yet clear but variants of both 326 and 343 are being frequently isolated in the southern part of the country (Luig, unpublished). It is interesting that in the 10 years since 326-1,2,3,5,6 was first isolated, its derivative, 343-1,2,3,5,6, has now become the most prevalent strain of *P. graminis tritici* in eastern Australia. The popularity of the cultivars Oxley, Condor, Egret and others which favour it has helped to change the population structure in this way.

As pointed out by Luig and Watson (1970), and more recently by Luig (1977), another branch in the evolutionary tree of the Australian stem rust flora had its origin in the production and establishment of strain 34-2,11. We have suggested that while the vast majority of Australian rusts have arisen as mutants, it is unlikely that 34-2,11 developed in this way. By simple mutation at the *P5* locus in the mid 1950's 34-2 arose from 21-0 and closely resembled it. Strain 34-2,11, however, differs from 21-0 in a number of genes (Luig and Watson, 1977). We believe that the source of this variation was somatic hybridization between races 21 and 126, occurring about 1957 when these were present in quantity on susceptible cultivars. This proposition is supported by further contrasting characteristics shown when the rusts are cultured on artificial media (Hartley and Williams, 1971).

Strain 34-2,11 underwent further variation to give 34-2,5,11, 34-2,4,5,11 and 40-2,4,5,11. None of the derivatives of this branch of the evolutionary tree has become important in the wheat areas except 34-2,4,5,11 which thrived during the period when Mendos (*Sr7a, Sr11, SrTt1, Sr17*) was grown. This may be because they failed to inherit the competitive ability of the race 21 parent.

From the simplified evolutionary diagram (Figure 9.2), four main developmental lines are recognizable. Two of them, that contributing the original six races and the one involved in the origin of the 126-222 group, are now extinct. Luig (1977) suggests diversity in the present rust flora stems from four basic strains 34-2,11, 21-0, 194-1,2,3,5,6 and 326-1,2,3,5,6. We can anticipate that future diversity will centre around the latter three groups and as new genes are utilized by the breeders, greater complexity in the strains will arise. In races 194 and 326 there will probably be as much variation as we have seen in the 21-34 group in the past.

It will be seen also from Figure 9.2 that during the 1960's and early 1970's strains virulent on plants with *SrTt1* were developing. Such rusts first became important when Mengavi was released. Strains 21-4,5, 21-2,4,5, 21-2,3,4,5,7, 34-2,4,11 and 34-2,4,5,11 arose progressively when Mendos replaced Mengavi. Although the gene for virulence on plants with *Sr6* is present in some cultures, we have failed to isolate from the field a strain in the 21-34 group of rusts with *p6 pTt1*. Consequently Timgalen, Songlen, Cook and Timson all having *Sr6* and *SrTt1* have remained resistant. We know the phenotype *p6 pTt1* can occur in the 126-222 group and that in the field mutation for *Sr6* and *SrTt1* will occur separately in the 21-34 group. Moreover the combined virulence has been induced in the glasshouse after E.M.S. treatment to give 34-1,2,3,4,5,6,7. Hence the north-eastern crop is currently quite vulnerable and crops of Oxley (*Sr6*) infected by 343-1,2,3,5,6 and growing adjacent to Cook or Songlen could provide the bridge necessary for an attack on these latter.

The present breeding programme

As will be seen from the account so far, breeding methodology has gone through evolutionary periods. Single genes were used from 1920 to 1950, the important ones being *Sr2, Sr6, Sr9b, Sr11* and *Sr17* and *SrTt1*. As evolution in the fungus proceeded, these and other genes were combined in various ways and the cultivars Mendos, Gamut, Timgalen, Gatcher, Songlen, Shortim, Oxley, Cook and Banks followed.

The fungus has been unable to match this approach by having appropriate virulence genes available in the right combinations at the right time. Hence losses from stem rust have been minimal. However, the method is complex and the synthesizing of combinations can only be done under controlled laboratory conditions. Moreover the effectiveness of the combinations must be monitored by a continuous survey of the rust populations in the field. Such surveys have had other advantages in that they have resulted in the isolation of new virulence genes or new combinations of them. This has enabled the genetic base of a cultivar to be broadened in advance of the cultivar being damaged in the field. Thus Timgalen was developed using a strain from the survey that had been isolated from Mendos. The genetic base in Songlen for resistance to *P. recondita* is being broadened by the use of a strain found attacking plants with *Lr17*.

When unavailable elsewhere, strains with particular virulence gene combinations may be produced by mutagenesis under controlled conditions, and used in the laboratory. In this way protection against anticipated strains can be accomplished in cultivars which already are resistant to all strains in the field. One such strain 34-1,2,3,4,5,6,7 which is virulent on seedlings of all current commercial cultivars with *Sr* genes, except those with *Sr26*, has been used to produce Songlen. Laboratory strains of this kind play a vital role in complementing the field nursery as a means of isolating new resistant material. To be successful, continuous genetic research on both host and fungus must be carried out.

The discovery and use of the gene *Sr26* has coincided with a reinstatement of single gene resistance breeding by some workers since it can be done economically in the field. *Sr26* is effective against all strains, and has remained so for the last 15 years; and while it would be desirable to combine other genes with *Sr26*, the derivation of wheat cultivars resistant to all current strains of *P. graminis tritici* can be easily accomplished using this valuable gene alone.

Because of the resources needed to breed for a broad base and because of the instability of most single gene resistances, many workers have urged that other methods of reducing the possibility of rust damage in wheat be examined. There have been suggestions for the use of multilines, cultivar mixtures, and of gene deployment in the different rust regions of the country (McIntosh, 1976). However no substantial work has so far encouraged breeders to implement these procedures and there appears to be no urgency to embark on such work at present. Preliminary studies, mainly of a pathological nature, have been done on a group of cultivars collectively known as "slow rusting" (Rees *et al.*, 1979). These workers developed the necessary skills in a series of experiments to establish those cultivars that allow only a slow rate of rust infection. So far the methods of integrating such data recording processes with a rapid and efficient selection procedure have not been devised, but these may come later. The greatest complication facing this work both in Australia and overseas is the problem of disentangling the effects of the well-known genes for specific resistance from the effects of those which could have an influence on the rate of rust infection. The genes *Sr6* and *Sr30* from Webster have presented problems, and those genes giving an intermediate type of resistance could also have a confounding effect. It could be, as Rees *et al.* (1979) suggest, that genes giving intermediate resistance, if effective in combination with others, could slow down still further the rate of rust infection. These genes of intermediate effect are doubtless specific, and to combine them presents a concept no different from the current idea of a resistance based on the combinations of the well-known *Sr* genes.

To establish irrefutably that a system of slow rusting, controlled independently of the currently known resistance genes is present in certain wheat cultivars, it will be necessary to remove the recognizable genes from the material. According to Rees *et al.* (1979) Gamut has slow rusting characters as well as at least four genes for specific resistance, and it was impossible to separate the effects in their experiments. One possible approach would be to cross Gamut and Federation, and continue backcrossing to Gamut, selecting for seedling susceptibility to a rust strain capable of eliminating the known genes. Should resulting lines without the specific resistance genes continue to be slow rusting, they could prove very useful as parents for subsequent programmes.

To summarize and to offer a prediction of how rust resistance breeding will proceed in the future, one has to say that in Australia success has resulted from three lines of approach all of which are aimed at achieving a more durable resistance.

(i) Parents have been selected on the basis of the diversity of their resistance genes and for a broad spectrum in their effectiveness against many strains. Data on their cytogenetical relationships has been available in gene catalogues (McIntosh, 1978) and where possible they have been combined in suitable agronomic backgrounds. Timgalen is a representative of new cultivars evolved in this way. Individual genes in the combinations may become ineffective and need to be replaced by other genes. Where the cultivar is highly resistant to all strains in the field and where further diversity in it may be restricted, it is necessary to search for strains which attack it. Alternatively, artificially induced mutants produced in the glasshouse which have additional virulences can be used in a controlled way in future studies. Songlen was produced from Timgalen in this way and replacements for Songlen are currently being assessed.

(ii) Use has been made of the gene *Sr2* conferring a durable resistance. This is effective after 50 years of use. Warigo was the first sucessful cultivar having it but others were produced later. This gene will be increasingly used in the future and research will be needed to increase yield in lines having this gene and to reduce the extent of damage due to head blackening associated with it.

(iii) Alien species have contributed to the success of the Australian programme and *A. elongatum* has been the most important donor in this respect. The gene *Sr26* will be more widely used in the future if it remains effective and if wheats to which it has been transferred maintain a high level of yield.

It can be anticipated that with the lessening of an immediate threat from stem rust more information will become available on leaf rust. As this occurs and as the genetic resources available to protect the crop against this disease are used, *P. recondita* will also be brought under control more effectively.

Acknowledgements

I am grateful to Professor B.D.H. Latter for making the facilities of the Plant Breeding Institute available to me, and to my colleagues Drs. R.A. McIntosh and N.H. Luig for useful discussions, for reading the manuscript and for allowing me to use their unpublished material.

References

de Sousa, C.N.A. (1975) A comparison of strains of *Puccinia graminis* f.sp. *tritici* from Angola and Rhodesia with their counterparts in Australia. M.Agr. Thesis, University of Sydney.

Flor, H.H. (1958) Mutation to wider virulence in *Melampsora lini. Phytopathology* 48, 297—301.

Hartley, M.J., and Williams, P.G. (1971) Morphological and cultural differences between races of *Puccinia graminis* f.sp. *tritici* in axenic culture. *Trans. Br. Mycol. Soc.* 57, 137—144.

Luig, N.H. (1977) The establishment and success of exotic strains of *Puccinia graminis tritici* in Australia. *Proc. Ecol. Soc. Aust.* 10, 89—96.

Luig, N.H. (1978) Mutation studies in *Puccinia graminis tritici.* Proc. 5th Int. Wheat Genetics Symp., New Delhi, 1978. Vol. 1, pp. 533—539.

Luig, N.H., and Watson, I.A. (1970) The effect of complex genetic resistance in wheat on the variability of *Puccinia graminis* f.sp. *tritici. Proc. Linn. Soc. N.S.W.* 95, 22—45.

Luig, N.H., and Watson, I.A. (1977) The role of barley, rye and grasses in the 1973-74 wheat stem rust epiphytotic in southern and eastern Australia. *Proc. Linn.. Soc. N.S.W.* 101, 65—76.

Maclean, D.J., Scott, K.J., and Tommerup, I.C. (1971) A uninucleate wheat-infecting strain of the stem rust fungus isolated from axenic cultures. *J. Gen. Microbiol.* 65, 339—342.

McIntosh, R.A. (1976) Genetics of wheat and wheat rusts since Farrer. *J. Aust. Inst. Agric. Sci.* 42, 203—216.

McIntosh, R.A. (1977) Nature of induced mutations affecting disease reaction in wheat. In 'Induced mutations against plant diseases'. International Atomic Energy Agency, Vienna. pp. 551—565.

McIntosh, R.A. (1978) A catalogue of gene symbols for wheat. Proc. 5th Int. Wheat Genetics Symp. New Delhi, 1978. Vol. 2, 1299—1309.

McIntosh, R.A., and Watson, I.A. (1981) Genetics of host-pathogen interaction in rusts. In 'The Rust Fungi'. Academic Press, London (in press).

Nelson, R.R., Wilcoxson, R.D., and Christensen, J.J. (1955) Heterocaryosis as a basis for variation in *Puccinia graminis* var. *tritici. Phytopathology* **45**, 639—643.

O'Brien, L., Brown, J.S., and Young, R.M. (1980) Occurrence and distribution of wheat stripe rust in Victoria and susceptibility of commercial wheat cultivars. *Aust. Plant Pathology* **9**, 14.

Rees, R.G., Thompson, J.P., and Goward, E.A. (1979) Slow rusting and tolerance to rusts in wheat. 1. The progress and effects of epidemics of *Puccinia graminis tritici* in selected wheat cultivars. *Aust. J. Agric. Res.* **30**, 403—419.

Waterhouse, W.L. (1936) Some observations on cereal rust problems in Australia. *Proc. Linn. Soc. N.S.W.* **61**, v—xxxviii.

Waterhouse, W.L. (1952) Australian rust studies. IX. Physiologic race determinations and surveys of cereal rusts. *Proc. Linn. Soc. N.S.W.* **77**, 209—258.

Watson, I.A. (1955) The occurrence of three new wheat stem rusts in Australia. *Proc. Linn. Soc. N.S.W.* **80**, 186—190.

Watson, I.A. (1957a) Mutation for pathogenicity in *Puccinia graminis* var. *tritici. Phytopathology* **47**, 507—509.

Watson, I.A. (1957b) Further studies on the production of new races from mixtures of races of *Puccinia graminis* var. *tritici* on wheat seedlings. *Phytopathology* **47**, 510—512.

Watson, I.A., and Luig, N.H. (1958) Somatic hybridization in *Puccinia graminis tritici. Proc. Linn. Soc. N.S.W.* **83**, 190—195.

Watson, I.A. and Luig, N.H. (1959) Somatic hybridisation between *Puccinia graminis* var. *tritici* and *Puccinia graminis* var. *secalis. Proc. Linn. Soc. N.S.W.* **84**, 207—208.

Watson, I.A., and Luig, N.H. (1963) The classification of *Puccinia graminis* var. *tritici* in relation to breeding resistant varieties. *Proc. Linn. Soc. N.S.W.* **88**, 235—258.

Watson, I.A., and Luig, N.H. (1966) *Sr15* — a new gene for use in the classification of *Puccinia graminis* var. *tritici. Euphytica* **15**, 239—250.

Watson, I.A., and Luig, N.H. (1968) Progressive increase in virulence in *Puccinia graminis tritici. Phytopathology* **58**, 70—73.

Williams, P.G. (1975) Evidence for diploidy of a monokaryotic strain of
  Puccinia graminis f.sp. *tritici*. *Trans. Br. Mycol. Soc.* **64**, 15–22.
Williams, P.G., and Mendgen, K.W. (1975) Cytofluorometry of DNA in uredo-
  spores of *Puccinia graminis* f.sp. *tritici*. *Trans. Br. Mycol. Soc.* **64**, 23–28.

10 WHEAT PROTEINS: THEIR CHEMISTRY AND NUTRITIONAL POTENTIAL

D.H. Simmonds

Introduction

The protein content of commercially grown wheat ranges from 6—16%. However, if an average figure of 10% is used, it may be calculated that the total amount of protein contributed by wheat to the human diet, is 43 Mt per annum. The composition of wheat proteins, and their supplementation with foods of complementary amino acid content, are therefore subjects of considerable importance in determining the nutritional status of millions of human beings.

Wheat has achieved its position as a basic food crop as a result of a number of attributes, not the least of which is the unique capacity of its protein to form an elastic, cohesive mass when mixed with water. Although most wheat varieties possess gluten forming capacity to a greater or lesser extent, the specific protein composition and properties which confer this capacity are still a matter for research, discussion and debate.

It is only quite recently (Johnson *et al.*, 1967; Shepherd, 1968; Wrigley, 1970; Wrigley and Shepherd, 1973; Orth and Bushuk, 1973, 1974; Bietz *et al.*, 1975; Lawrence and Shepherd, 1980) that the location of genes coding for the synthesis of specific wheat proteins has been established. This approach provides some sound basis for the classification of wheat proteins, a problem which has plagued cereal chemists for decades. This chapter will summarize some aspects of the chemistry and structure of wheat proteins, and discuss this in relation to their properties and nutritional potential.

The chemistry of wheat proteins

One of the reasons why the relationship between the composition of wheat proteins and their physical properties has remained obscure in spite of the efforts of research workers ranging over nearly two centuries is the complexity of this group of substances. The results of these efforts have shown that the storage protein fraction of wheat comprises several families of closely related substances, which appear to have arisen as a result of gene mutation, duplication, and recombination. The extent to which selection pressure has operated to allow multiple forms of these proteins to persist in modern bread wheats is still unclear. It is possible that the possession of a wide spectrum of closely related storage proteins confers some specific advantages on the plant, although the extent to which this is operative in a widely adapted, cultivated crop species would be difficult to demonstrate.

According to the present status of knowledge, wheat proteins may be classified into two broad groups — the cytoplasmic or metabolically active proteins, and the storage proteins. The former comprise most of the albumins and globulins of the classical literature, while the latter include the less soluble gliadins and glutenins first characterized in detail by Osborne (1907).

Like all generalizations, such a classification tends to obscure important details. The mature wheat kernel is a complex entity comprising both living (respiring) and non-living portions. Even the non-respiring portions, the endosperm and pericarp, were metabolically active during the development of the kernel, and only became non-living as maturity approached. All cells comprising the caryopsis therefore have, or have possessed at some time, a full complement of organelles, membranes, synthetic and degradative enzymes. These, or their remnants are present at maturity, and are carried through into wheat-derived products during milling and further processing (Simmonds, 1972 a,b; Simmonds and Wrigley, 1972).

Cytoplasmic proteins

Cytoplasmic proteins in general are water-, buffer-, or salt-soluble, and comprise numerous enzymes (e.g., Preston and Kruger, 1976, 1977; Kruger, 1972, 1973, 1976, 1977), as well as enzyme inhibitors capable of interfering with the activity of amylase and protease enzymes in other species (Buonocore *et al.*, 1977). The latter are thought to confer resistance to insect attack.

Fractionation of the cytoplasmic proteins shows them to be a complex mixture of substances which can best be resolved by two dimensional electrofocusing/electrophoresis. Such a separation is illustrated in Figure 10.1.

Storage proteins

The storage proteins of the wheat kernel are laid down within the cells of the developing endosperm in specialized organelles termed protein bodies (Graham *et al.*, 1962; Jennings *et al.*, 1963). These first make their appearance some 10-12 days after flowering, and develop to fill the vacuolar space within the cell. Eventually they increase in size and become compressed between the starch granules of the mature endosperm.

The manner in which the storage proteins are synthesised and transported into the vacuoles has not yet been fully resolved, although descriptions of the process have been given by several authors (Buttrose, 1963; Graham *et al.*,

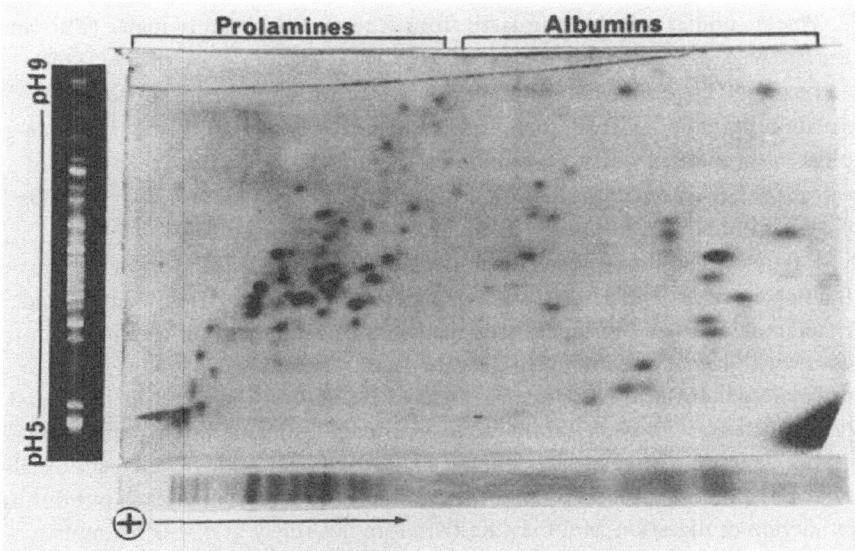

Fig. 10.1. Two dimensional electrofocusing/electrophoresis of the gliadins and the buffer-soluble proteins isolated from wheat flour. (Wrigley, 1976).

1963a; Jennings *et al.*, 1963; Jennings and Morton, 1963; Barlow *et al.*, 1974; Simmonds, 1978; Briarty *et al.*, 1979).

The most recent evidence suggests that synthesis takes place principally on membrane-bound polyribosomes, and that the protein thus synthesised accumulates within the lumen of the endoplasmic reticulum, whence it is transported into the vacuoles. Whether the vacuoles are formed from the cisternae of the endoplasmic reticulum as suggested by Briarty *et al.* (1979) is unresolved at this time. This process has been documented in maize by Burr and Burr (1976) and by Khoo and Wolf (1970). However, in the case of wheat (Buttrose, 1963) and in legumes (Dieckert and Dieckert, 1972) the intermediate role of dictyosomes has been suggested.

Isolation and characterization of protein bodies from developing wheat endosperm was first achieved by Graham *et al.* (1963b). These workers showed that the principal proteins present were acetic acid-soluble, and had the electrophoretic mobilities characteristic of the gliadin and glutenin groups. Yields were very low — only about 5% of the total protein bodies present in the intact endosperm were recovered, and these comprised mainly the smaller sized organelles ranging in diameter between 0.1 and 1.2 micrometers.

Protein bodies have been isolated from other cereals notably maize (Burr and Burr, 1976), sorghum (Adams *et al.*, 1976), rice (Mitsuda *et al.*, 1967, 1969); and barley (Ory and Henningsen, 1969). It has proved to be less difficult to isolate protein bodies from these sources than from wheat. Possibly the same physical properties which contribute to the unique rheological behaviour of wheat proteins are also responsible for the difficulties encountered in isolating protein bodies from this cereal in an intact and undamaged form.

Reference has already been made to the complexity of the storage protein fraction in wheat. The difficulties inherent in extracting, fractionating and characterizing these substances are reflected in the number of different techniques which have been employed over the years. The storage proteins of the wheat kernel occur in two principal areas of the grain — the aleurone layer, and the endosperm. Those of the aleurone are present as membrane-bounded inclusions enclosed within a tough, chemically resistant cell wall. These cells usually survive the milling process intact; their contents do not readily leach out during extraction or digestion, and they are therefore not freely available for human (and other animal) nutrition.

Investigation of the endosperm storage proteins presents problems because of their unusual solubility characteristics. These proteins differ from the cytoplasmic group in being essentially insoluble in water and dilute aqueous salt solutions, although some of the gliadin group proteins dissolve partly in cold, and more readily in hot water. This is due to their unusual amino acid composition (see later) which is exceptionally low in charged and polar groups capable of interacting with an aqueous environment.

Most of the solvents which have been employed, are of a dissociating nature, attacking either the covalent, electrostatic, hydrophobic, or hydrogen bonds responsible for causing the individual protein species to aggregate. The amino acid residues involved in the four types of interaction are listed in Table 10.1. The various reagents used for dissociation and solubilization react with the amino acid side chains according to their chemical nature, and the conformation of the protein in its original environment must be deduced from a knowledge of the specific effects mediated by the solvent.

Characterization of wheat storage proteins

Once dispersed in solution, wheat proteins may be fractionated and characterized by the various techniques which have been developed by protein chemists. Gel electrophoresis and electrofocusing, either alone or in combination have been shown to yield the highest resolving power currently obtainable (Wrigley, 1976; du Cros and Wrigley, 1979).

Table 10.1

Grouping of amino acid residues according to their interactions within and between protein chains

1. Electrostatic interactions	(i) Acid hydrophilic:
Dissociated by acid or alkali, salt solutions	Aspartic acid Glutamic acid
	(ii) Basic hydrophilic:
	Lysine Arginine Histidine
2. Neutral — hydrogen bonding	Asparagine Glutamine
Dissociated by urea, dimethyl formamide	Threonine Serine Cysteine
3. Neutral — hydrophobic interaction	Tyrosine Tryptophan
Dissociated by ionic and non-ionic detergents, sodium salts of long-chain fatty acids	Phenylalanine Proline Methionine Leucine Isoleucine Valine Alanine Glycine
4. Covalent — disulphide bonding	Cystine
Dissociated by reducing agents, e.g., 2-mercaptoethanol	

Starch gel (Smithies, 1955) was the original medium used during zonal electrophoresis to support the sample and electrolyte (Woychik et al., 1961, 1964; Elton and Ewart, 1962; Graham, 1963; Wrigley and Shepherd, 1974). However, its place has been largely taken by acrylamide gels having either uniform pore size (Wrigley, 1976) or a gradient in cross-linking concentration (Margolis and Wrigley, 1975; du Cros and Wrigley, 1979). Procedures for both acrylamide gel electrophoresis and electrofocusing have been described in detail by Wrigley (1976) and McCausland and Wrigley (1977). A further modification of the acrylamide gel electrophoresis technique, has been the use of SDS as a dissociating medium both for the extraction of protein and during electrophoresis (Weber and Osborn, 1969; Bietz and Wall, 1972; Orth and Bushuk, 1973, 1974). As a result, the behaviour of wheat proteins, both in the oxidised ($-S-S-$) and reduced ($-SH$; sub-unit) forms under a variety of electrophoretic conditons has been well documented.

Figure 10.2 shows the starch gel electrophoresis patterns given by the gliadin fraction of a number of Australian wheat varieties, while Figure 10.3 shows the separation achieved using a gradient pore acrylamide gel as the supporting medium (du Cros and Wrigley, 1979). The greater resolution attainable in the latter system is immediately apparent.

Close inspection of Figures 10.2 and 10.3 shows that the gliadin group of proteins from each cultivar gives a clearly distinguishable and characteristic pattern. These patterns are a function of grain phenotype, and remain characteristic of the cultivar, regardless of the environmental conditions, with one notable exception which has been recognized so far. As a result, electrophoresis of gliadin proteins has now been developed as a widely used method of variety identification (Autran et al., 1979).

The notable exception, is where the grain has been grown under conditions of severe sulfur deficiency (Wrigley et al., 1980). Under these conditions an increase is observed in the proportion of electrophoretically slow-moving (low mobility) gliadins present in the patterns. This group of proteins, whose synthesis is controlled by genes on group-1 chromosomes (Wrigley, 1970; Wrigley and Shepherd, 1973), appear to be virtually devoid of sulfur-containing amino acids, and their synthesis under conditions of severe sulfur deficiency is less affected than that of the high-mobility, high-sulfur group. The amounts of albumin/globulin proteins synthesised are also reduced under conditions of sulfur deficiency. These changes in protein composition affect both the quality characteristics and amino acid composition of flours prepared from sulfur deficient wheats. The low-sulfur grain is severely depleted in all nutritionally essential amino acids, while increases are observed in the contents of arginine and aspartic acid (Wrigley et al., 1980).

OLYMPIC

GLENWARI

HOPPS

GLUCLUB

INSIGNIA

WREN

WINDEBRI

DURAL

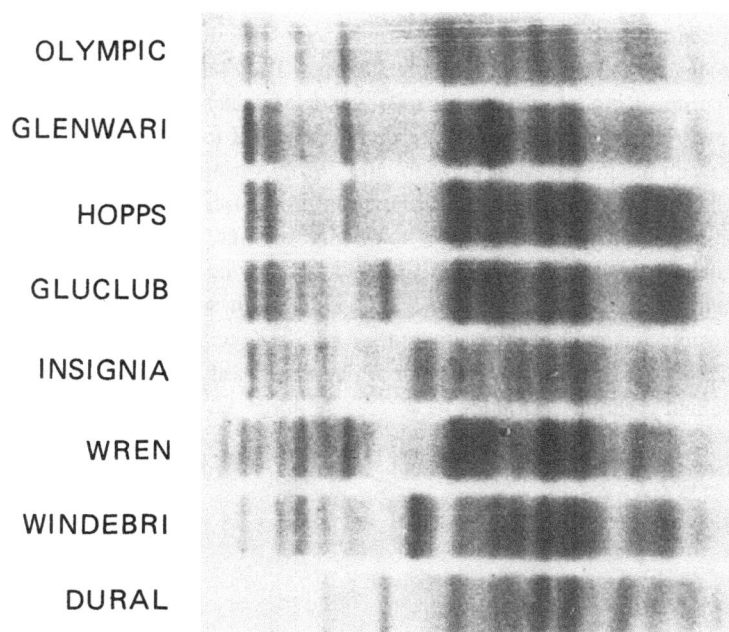

Fig. 10.2. Starch gel electrophoresis of wheat gliadin proteins in aluminium lactate buffer of pH 3.2 by the method of Graham (1963).

GAMENYA B
GLUCLUB
GATCHER
HERON
BINDAWARRA
AVOCET
BANKS
WARIGAL
HALBERD
MILLEWA
TINCURRIN
MILING

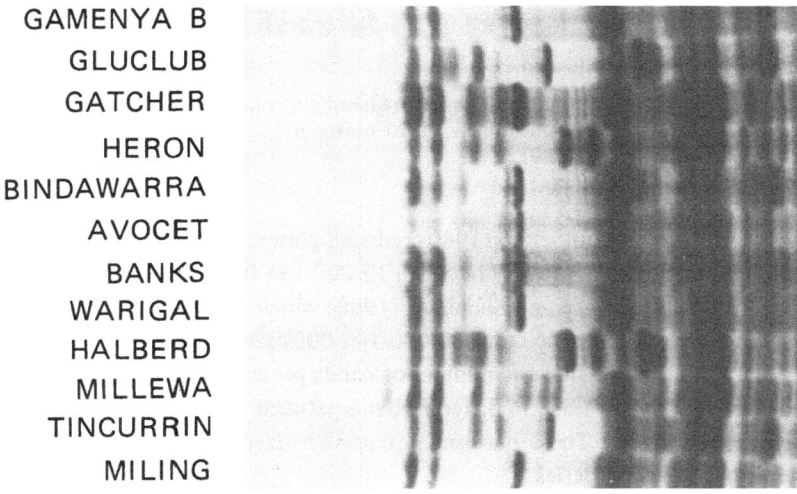

Fig. 10.3. Gradient acrylamide gel electrophoresis in sodium lactate buffer, of endosperm proteins from Australian wheat varieties (du Cros and Wrigley, 1979).

The albumin and globulin groups of proteins are electrophoretically faster moving on starch gel than are the gliadins. They have not been observed to vary in composition between cultivars of a single species. They do however, appear to vary between species, and are therefore useful for distinguishing, for instance, between bread wheats as a group, and durum varieties (Wrigley, 1977).

In the electrophoretic patterns shown in Figures 10.2 and 10.3, the glutenin (high MW) proteins have either not been extracted, or if present, have largely remained in the starting slots of the gel. Following reduction with 2-mercapto-ethanol and dissociation with SDS, the glutenin sub-units are brought into solution and may be fractionated by SDS-polyacrylamide gel electrophoresis (SDS-PAGE) on the basis of their molecular size alone. The type of separation attained is shown in Figure 10.4.

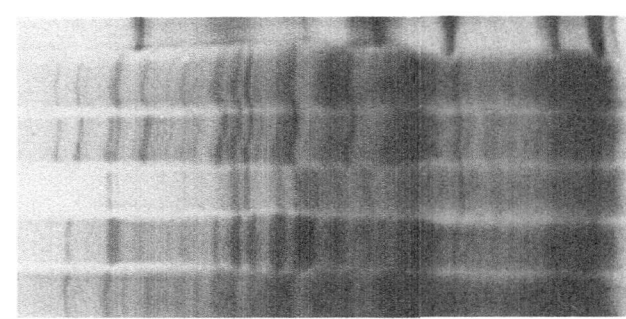

Fig. 10.4. SDS-electrophoresis of reduced glutenin from four varieties of wheat. Separations performed in phosphate buffer, pH 7.2, containing 0.1% SDS. (Wrigley *et al.*, 1980).

According to Bushuk *et al.* (1980) reduced glutenin comprises approximately 17 sub-units, ranging in MW between 12 000-134 000. These results differ from those of Lawrence and Shepherd (1980) which suggest that only the sub-units of higher MW (in the range 80 000-140 000) are derived from glutenin. Their studies showed that the number of bands present ranged between three and five, although at least 34 different band patterns were observed in the 98 cultivars examined. These patterns potentially offer an alternative means of identifying wheat varieties.

Formation and structure of wheat gluten

Wheat gluten as such does not exist in the mature wheat kernel, nor in flour derived from wheat endosperm. Gluten is only produced when flour is mixed with water and subjected to a small input of mechanical work. An essential difference between gluten and the storage protein from which it is derived, is that during mixing with water, lipids from endosperm cell membranes have reacted with the storage protein to produce lipid-protein interaction products of some complexity (Simmonds and Wrigley, 1972). These are considerably less soluble than the storage proteins from which they were derived, and contribute significantly to the high molecular, glutenin fraction.

Wheat gluten is therefore a complex in which the original gliadin and glutenin fractions of the storage protein have been combined with membrane-derived lipid. A close association has also been demonstrated between a carbohydrate and some of the high MW glutenin sub-units (Khan and Bushuk, 1979; Bushuk *et al.*. 1980). Preliminary results suggest that the linkage between the carbohydrate and the protein is a covalent one, so that the resulting glycoprotein appears to be involved in the association of specific glutenin sub-units into large aggregates. The lipid-protein-carbohydrate complex so formed confers upon wheat flour doughs, their exceptional rheological characteristics. In this complex, the glutenin fraction appears to play the more important role, both in providing a basic framework around which the rest of the complex is built, as well as determining its rheological properties, and hence its quality character-istics. It is the latter which make wheat flour valuable for a wide variety of uses, and which are responsible for the widespread consumption of wheat throughout the world. In the model proposed by Bushuk and co-workers most of the glutenin sub-units are linked into a central disulphide-crosslinked matrix by secondary forces (hydrogen bonds, hydrophobic and ionic interactions). The central matrix is composed mainly of the high MW sub-units, and plays a major role in determining the suitability of wheat flour doughs for breadmaking.

Nutritional properties of wheat proteins

All the morphological parts of the wheat grain contain protein, with the embryo and scutellum containing the highest concentrations per unit weight. However, as shown in Figure 10.5, because of their small size, these compon-ents contribute very little to the total protein of the grain.

The major proportion of the total protein (usually between 75-85%), is contributed by the gliadin and glutenin components of the storage protein. These components therefore strongly influence the amino acid composition of wheat flour. This is shown in Table 10.2 in which the amino acid compositions of the albumin, globulin, gliadin and glutenin fractions are compared with that of whole wheat flour.

Table 10.2

The amino acid composition of the albumin, globulin, gliadin and glutenin fractions of wheat flour proteins. (g amino acid/16 gN)

Amino Acid	Whole[a] wheat	Wheat[b] flour	Albumin[c]	Globulin[b]	Gliadin[d]	Glutenin[d]	Whole gluten[e]
Alanine	3.6	2.6	4.64	4.9	1.85	2.53	2.26
Amide	3.5	3.9	0.41	1.9	4.66	3.68	3.74
Arginine	4.6	3.1	7.12	8.3	2.51	3.82	3.66
Aspartic acid	4.9	3.7	7.11	7.0	2.60	3.41	3.92
½ Cysteine	2.5	2.8	5.98	5.4	2.70	2.16	2.10
Glutamic acid	29.9	34.7	16.23	15.5	35.45	30.24	32.11
Glycine	4.1	3.4	2.47	4.9	1.40	3.49	3.31
Histidine	2.3	1.9	3.95	2.6	2.04	2.11	2.24
Isoleucine	3.3	3.1	3.72	3.2	3.92	3.41	3.55
Leucine	6.7	6.6	9.68	6.8	6.26	6.00	6.77
Lysine	2.9	1.9	10.05	5.9	0.58	2.06	1.82
Methionine	1.5	1.3	0.00	1.7	1.30	1.51	1.76
Phenylalanine	4.5	4.8	4.70	3.5	5.03	4.35	6.12
Proline	9.9	11.8	7.41	5.0	12.57	9.38	11.55
Serine	4.6	4.4	4.03	4.8	4.26	4.91	4.35
Threonine	2.9	2.4	2.53	3.3	1.96	2.83	2.50
Tryptophan	1.1	1.5	n.d.	1.1	0.65	1.90	n.d.
Tyrosine	3.0	2.8	3.17	2.9	2.38	3.29	3.65
Valine	4.4	3.4	7.19	4.6	3.76	3.87	3.63

a. FAO (1970); b. Bushuk & Wrigley (1974); c. Waldschmidt-Leitz & Hochstrasser (1961); d. Ewart (1967); e. Batey (personal communication).

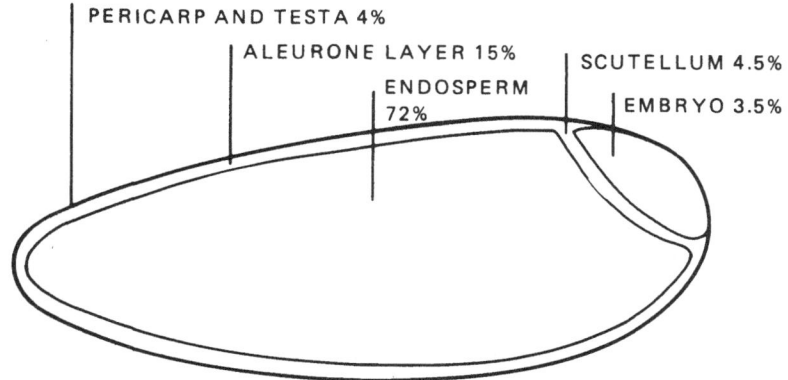

PERICARP AND TESTA 4%

ALEURONE LAYER 15%

ENDOSPERM 72%

SCUTELLUM 4.5%

EMBRYO 3.5%

Fig. 10.5. Distribution of protein within the wheat kernel (based on MacMasters *et al.*, 1971).

The gliadin fraction, which comprises some 40-50% of the total flour protein, is notably deficient in the essential amino acid lysine, and to a lesser extent in threonine, compared with the FAO/WHO reference protein derived from estimates of human amino acid requirements (FAO/WHO, 1973; Saxelby and Venn-Brown, 1980). The other protein fractions contain insufficient of these essential amino acids to compensate for these deficiencies, and so wheat flour as a whole, and more particularly gluten and gluten-derived products, would appear to require fortification or balancing with other lysine-containing foodstuffs in order to provide a diet which is nutritionally ideal. In spite of this, experiments by several research groups (e.g. Edwards *et al.*, 1971) have shown that the proteins of wheat alone can satisfy the amino acid requirements of the human adult, provided sufficient quantities are eaten. The lysine requirement for an adult is 700-800 mg/day (FAO/WHO, 1973). This can be met by the consumption of 400 g (13-15 slices) of white bread per day, which at the same time supplies 4120 KJ, or about 43% of the adult daily energy requirements (Saxelby and Venn-Brown, 1980).

In the case of growing children, nursing mothers, and other special groups however, the evidence is less conclusive. In most cases these groups could not consume the bulk of material needed to supply their daily requirements, and their diet must be supplemented by a more concentrated protein source.

Attempts have been made to raise the lysine contents of bread wheats by selective breeding (Johnson et al., 1968), and various wheat collections have been screened (Mattern et al., 1970; Lehmann et al., 1978) in an attempt to obtain suitable parental material. Unfortunately, high lysine mutants have not yet been identified in wheat, and other means which have been tried, such as the artificial enrichment of wheat by the addition of lysine (Ferrel et al., 1970; Kato and Muramatsu, 1971), have not always received wide acceptance.

Bread formulations which are artificially enriched in protein of non-fat milk solids, soy flour, and yeast have higher contents of essential amino acids than breads containing wheat flour alone. These additions also help to offset the losses of lysine, estimated at between 10-20%, which are caused by browning or Maillard reactions during baking (Rosenberg and Rohdenburg, 1951).

Except for particularly disadvantaged groups, the reduced levels of lysine and threonine in wheat proteins do not generally pose a problem in Australia because of our varied diet and high intake of animal protein. However, for a variety of social, economic and political reasons, many overseas countries have increased their production and importation of wheat, and are becoming increasingly dependent upon this cereal as a basic source of protein and calories. In the long term this may pose problems.

One way of meeting these problems lies in the development of nutritionally balanced, organoleptically-attractive basic protein products from alternative food sources. Such products are likely to be derived from vegetable proteins in view of the high cost and restricted availability of animal protein. In its position as a major cereal crop, and because of the specific solubility and rheological properties of its storage proteins, wheat is in a favourable position to provide the basic raw material from which some of these protein products may be prepared.

A second and potentially important reason for selecting wheat for this purpose in Australia is the possible use of this crop as a source of starch for fuel ethanol production (Stewart et al., 1979). It has been estimated (Simmonds, 1980) that the starch derived from 1 Mt of wheat would provide 300 000 tonnes of ethanol, sufficient to treat 3 Mt of petrol at the 10% level, or roughly one fifth of our present national consumption. The by-product protein from this operation would amount to approximately 100 000 tonnes, and this material therefore represents a raw material of considerable potential value. In addition, if the production of fuel ethanol by fermentation of cereal carbohydrate became at all widely established, it is likely that significant quantities of microbial protein would also be produced.

Table 10.3

The essential amino acid composition of wheat gluten, yeast protein, and a 50:50 mixture of these, compared with beef, rice, and the FAO/WHO reference protein (mg amino acid/g Nitrogen) (FAO, 1970; Saxelby and Venn Brown, 1980).

Amino acid	Rice	Beef	Wheat Gluten	Yeast Protein	Wheat +Yeast	WHO/FAO Reference Protein
Isoleucine	240	300	267	310	288	250
Leucine	512	506	472	450	460	440
Lysine	238	556	121	510	316	340
Methionine	144	170	95	110	102	
Cystine	70	80	130	60	95	
Met + Cys	212	250	225	170	197	220
Phenylalanine	325	275	364	270	317	
Tyrosine	220	225	248	230	239	
Phe + Tyr	544	500	612	500	556	380
Threonine	244	288	167	300	234	250
Tryptophan	80	70	70	70	70	60
Valine	344	312	260	400	330	310

As shown in Table 10.3, the amino acid compositions of wheat protein and yeast protein are in many ways complementary, so that suitable blends could be prepared, having a nutritionally improved amino acid composition.

The suitability of such blends for a variety of food uses both within Australia and overseas has yet to be demonstrated. However, such developments are of potential importance to those areas of the world, especially in the tropics, where starchy food crops can readily be grown, but where protein is less freely available. In the case of these regions it may make good economic sense to prepare a basic protein concentrate of sound nutritional value for export from countries like Australia, rather than to export a bulky raw material such as wheat which is already rich in starch.

Apart from their potential as a basic protein raw material for human consumption, wheat proteins also have wide application in the formulation of pet foods, and indeed animal feeds generally. This is a growing market in S.E. Asia where government policies are aimed at increasing the available supplies of animal protein.

Many of the problems faced by developing nations in securing adequate food supplies for their growing populations are economic, cultural, and social. In spite of this, important areas of research and development remain for the food technologist, both now and in the future.

Acknowledgements

It is a pleasure to acknowledge the valuable discussions held with Dr. C.W. Wrigley who also directed most of the work concerned with the electrophoretic and electrofocusing separations. Mrs. D.L. du Cros and Mrs. P.A. Boniface were responsible for the preparation of the figures.

References

Adams, C.A., Novellie, L., and Liebenberg, N.V. de W. (1976) Biochemical properties and ultrastructure of protein bodies isolated from selected cereals. *Cereal Chem.* **53**, 1–12.

Autran, J.C., Bushuk, W., Wrigley, C.W., and Zillmann, R.R. (1979) Wheat cultivar identification by gliadin electrophoregrams. IV. Comparison of international methods. *Cereal Foods World* **24**, 471–475.

Barlow, K.K., Lee, J.W., and Vesk, M. (1974) Morphological development of storage protein bodies in wheat. In 'Mechanisms of Regulation of Plant Growth'. (Eds. R.L. Bieleski, A.R. Ferguson, and M.M. Cresswell.) Royal Soc. New Zealand Bull. No. 12, 793–797.

Bietz, J.A., and Wall, J.S. (1972) Wheat gluten sub-units: Molecular weights determined by sodium dodecyl sulfate – polyacrylamide gel electrophoresis. *Cereal Chem.* **49**, 416–430.

Bietz, J.A., Shepherd, K.W., and Wall, J.S. (1975) Single-kernel analysis of glutenin : use in wheat genetics and breeding. *Cereal Chem.* **52**, 513–532.

Briarty, L.G., Hughes, C.E., and Evers, A.D. (1979) The developing endosperm of wheat – a stereological analysis. *Ann. Bot. (Lond.)* **44**, 641–658.

Buonocore, V., Petrucci, T., and Silano, V. (1977) Wheat protein inhibitors of alpha-amylase. *Phytochemistry* **16**, 811–820.

Burr, B., and Burr, F.A. (1976) Zein synthesis in maize endosperm by polyribosomes attached to protein bodies. *Proc. Natl. Acad. Sci. U.S.A.* **73**, 515–519.

Bushuk, W., and Wrigley, C.W. (1974) Proteins: Composition, structure and function. In 'Wheat: Production and Utilization'. (Ed. G.E. Inglett.) AVI Publishing Co. Inc, Westport, Conn. pp. 119–145.

Bushuk, W., Khan, K., and McMaster, G. (1980) Functional glutenin: Complex of covalently and noncovalently linked components. Proc. workshop on the physico-chemical properties of wheat gluten proteins, Nantes, France, April 28–30, 1980 (in press).

Buttrose, M.S. (1963) Ultrastructure of the developing wheat endosperm. *Aust. J. Biol. Sci.* **16**, 305–317.

Dieckert, J.W., and Dieckert, M.C. (1972) The deposition of vacuolar proteins in oilseeds. In 'Symposium: Seed Proteins'. (Ed. G.E. Inglett.) AVI Publishing Co. Inc, Westport, Conn. pp. 52–85.

du Cros, D.L., and Wrigley, C.W. (1979) Improved electrophoretic methods for identifying cereal varieties. *J. Sci. Food Agric.* **30**, 785–794.

Edwards, C.H., Booker, L.K., Rumph, C.H., and Ganapathy, S.N. (1971) Utilization of wheat by adult man: excretion of vitamins and minerals. *Am. J. Clin. Nutr.* **24**, 547–555.

Elton, G.A.H., and Ewart, J.A.D. (1962) Starch-gel electrophoresis of cereal proteins. *J. Sci. Food Agric.* **13**, 62–72.

Ewart, J.A.D. (1967) Amino acid analysis of glutenins and gliadins. *J. Sci. Food Agric.* **18**, 111–116.

FAO (1970) Amino acid content of foods and biological data on proteins. FAO Nutritional studies No. 24. Food and Agriculture Organization of the United Nations, Rome.

FAO/WHO (1973) Energy and protein requirements. Report of a Joint FAO/WHO Ad Hoc Expert Committee. World Health Organization Tech. Rep. Ser. No. 522; Food and Agriculture Organization Nutr. Meetings Rep. Ser. No. 52. Geneva.

Ferrel, R.E., Shepherd, A.D., and Guadagni, D.G. (1970) Storage stability of lysine-fortified wheat. *Cereal Chem.* **47**, 33–37.

Graham, J.S.D. (1963) Starch gel electrophoresis of wheat flour proteins. *Aust. J. Biol. Sci.* **16**, 342–349.

Graham, J.S.D., Jennings, A.C., Morton, R.K., Palk, B.A., and Raison, J.K. (1962) Protein bodies and protein synthesis in developing wheat endosperm. *Nature* **196**, 967–969.

Graham, J.S.D., Morton, R.K., and Simmonds, D.H. (1963a) Studies of proteins of developing wheat endosperm. Fractionation by ion-exchange chromatography. *Aust. J. Biol. Sci.* **16**, 350–356.

Graham, J.S.D., Morton, R.K., and Raison, J.K. (1963b) Isolation and characterization of protein bodies from developing wheat endosperm. *Aust. J. Biol. Sci.* **16**, 375–383.

Jennings, A.C., and Morton, R.K. (1963) Amino acids and protein synthesis in developing wheat endosperm. *Aust. J. Biol. Sci.* **16**, 384–394.

Jennings, A.C., Morton, R.K., and Palk, B.A. (1963) Cytological studies of protein bodies of developing wheat endosperm. *Aust. J. Biol. Sci.* **16**, 366–374.

Johnson, B.L., Barnhart, D., and Hall, O. (1967) Analysis of genome and species relationships in polyploid wheats by protein electrophoresis. *Am. J. Bot.* **54**, 1089–1098.

Johnson, V.A., Schmidt, J.W., and Mattern, P.J. (1968) Cereal breeding for better protein impact. *Econ. Bot.* **22**, 16–25.

Kato, J., and Muramatsu, N. (1971) Amino acid supplementation of grain. *J. Am. Oil Chem. Soc.* **48**, 415–419.

Khan, K. and Bushuk, W. (1979) Studies of glutenin. XIII. Gel filtration, iso-electric focusing and amino acid composition studies. *Cereal Chem.* **56**, 505–512.

Khoo, U., and Wolf, M.J. (1970) Origin and development of protein granules in maize endosperm. *Am. J. Bot.* **57**, 1042–1050.

Kruger, J.E. (1972) Changes in the amylases of hard red spring wheat during growth and maturation. *Cereal Chem.* **49**, 379–390.

Kruger, J.E. (1973) Changes in the levels of proteolytic enzymes from hard red spring wheat during growth and maturation. *Cereal Chem.* **50**, 122–131.

Kruger, J.E. (1976) Changes in the polyphenol oxidases of wheat during kernel growth and maturation. *Cereal Chem.* **53**, 201–213.

Kruger, J.E. (1977) Changes in the catalases of wheat during kernel growth and maturation. *Cereal Chem.* **54**, 820–826.

Lawrence, G.J., and Shepherd, K.W. (1980) Variations in glutenin sub-units of wheat. *Aust. J. Biol. Sci.* **33**, 221–233.

Lehmann, C.O., Rudolph, A., Hammer, K., Meister, A., Muntz, K., and Scholz, F. (1978) Protein screening on the Gatersleben collections of cereals and pulses. 1. Protein and lysine content of wheats and their interspecific and intergeneric hybrids. *Kulturpflanze* **26**, 133–161.

MacMasters, M.M., Hinton, J.J.C. and Bradbury, D. (1971) Microscopic structure and composition of the wheat kernel. In 'Wheat: Chemistry and Technology'. (Ed. Y. Pomeranz.) American Association of Cereal Chemists, St. Paul, Minn. pp. 51–113.

McCausland, J., and Wrigley, C.W. (1977) Identification of Australian barley cultivars by laboratory methods: gel electrophoresis and gel isoelectric focusing of the endosperm proteins. *Aust. J. Exp. Agric. Anim. Husb.* **17**, 1020–1027.

Margolis, J., and Wrigley, C.W. (1975) Improvement of pore gradient electrophoresis by increasing the degree of cross-linking at high acrylamide concentrations. *J. Chromatogr.* **106**, 204–209.

Mattern, P.J., Schmidt, J.W., and Johnson, V.A. (1970) Screening for high lysine content in wheat. *Cereal Sci. Today.* **15**, 409–411.

Mitsuda, H., Yasumoto, K., Murakami, K., Kusano, T., and Kishida, H. (1967) Studies on the proteinaceous subcellular particles in rice endosperm: electron microscopy and isolation. *Agric. Biol. Chem.* **31**, 293–300.

Mitsuda, H., Murakami, K., Kusano, T., and Yasumoto, K. (1969) Fine structure of protein bodies isolated from rice endosperm. *Arch. Biochem. Biophys.* **130**, 678–680.

Orth, R.A., and Bushuk, W. (1973) Studies of glutenin. III. Identification of sub-units coded by the D-genome and their relation to breadmaking quality. *Cereal Chem.* **50**, 680–687.

Orth, R.A., and Bushuk, W. (1974) Studies of glutenin. IV. Chromosomal location of genes coding for subunits of glutenin of common wheat. *Cereal Chem.* **51**, 118–126.

Ory, R.L., and Henningsen, K.W. (1969) Enzymes associated with protein bodies isolated from ungerminated barley seeds. *Plant Physiol.* **44**, 1488–1498.

Osborne, T.B. (1907) Proteins of the wheat kernel. Carnegie Inst., Washington. Publ. 84.

Preston, K.R., and Kruger, J.E. (1976) Purification and properties of two proteolytic enzymes with carboxypeptidase activity in wheat. *Plant Physiol.* **58**, 516–520.

Preston, K., and Kruger, J.E. (1977) Specificity of two isolated wheat carboxy-peptidases. *Phytochemistry* **16**, 525–528.

Rosenberg, H.R., and Rohdenburg, E.L. (1951) The fortification of bread with lysine. 1. The loss of lysine during baking. *J. Nutr.* **45**, 593–598.

Saxelby, C., and Venn-Brown, U. (1980) 'The Role of Australian Flour and Bread in Health and Nutrition.' Bread Research Institute of Australia, Sydney.

Shepherd, K.W. (1968) Chromosomal control of endosperm proteins in wheat and rye. Proc. 3rd Int. Wheat Genetics Symp. Canberra, Australia. pp. 86–96.

Simmonds, D.H. (1972a) The ultrastructure of the mature wheat endosperm. *Cereal Chem.* **49**, 212–222.

Simmonds, D.H. (1972b) Wheat grain morphology and its relationship to dough structure. *Cereal Chem.* **49**, 324–335.

Simmonds, D.H. (1978) Structure, composition and biochemistry of cereal grains. In 'Cereals '78: Better Nutrition for the World's Millions'. (Ed. Y. Pomeranz.) American Association of Cereal Chemists Inc., St. Paul, Minn. pp. 105–137.

Simmonds, D.H. (1980) The Australian cereal crop and its potential for the production of fuel and food. *Food Technol. Aust.* **32**, 442–446.

Simmonds, D.H., and Wrigley, C.W. (1972) The effect of lipid on the solubility and molecular weight range of wheat gluten and storage protein. *Cereal Chem.* **49**, 317–323.

Smithies, O. (1955) Zone electrophoresis in starch gels: Group variations in the serum proteins of normal human adults. *Biochem. J.* **61**, 629–641.

Stewart, G.A., Gartside, G., Gifford, R.M., Nix, H.A., Rawlins, W.H.M. and Siemon, J.R. (1979) 'The Potential for Liquid Fuels from Agriculture and Forestry in Australia.' Commonwealth Scientific and Industrial Research Organization, Melbourne.

Waldschmidt-Leitz, E. and Hochstrasser, K. (1961) Uber die Albumine aus Gerste und Weizen. VII. Mitteilung uber Samenproteine. *Hoppe-Seyler's. Z. Physiol. Chem.* **324**, 243.

Weber, K., and Osborn, M. (1969) The reliability of molecular weight determinations by dodecyl sulfate-polyacrylamide gel electrophoresis. *J. Biol. Chem.* **244**, 4406–4412.

Woychik, J.H., Boundy, J.A., and Dimler, R.J. (1961) Starch-gel electrophoresis of wheat gluten proteins with concentrated urea. *Arch. Biochem. Biophys.* **94**, 477–482.

Woychik, J.H., Huebner, F.R., and Dimler, R.J. (1964) Reduction and starch gel electrophoresis of wheat gliadin and glutenin. *Arch. Biochim. Biophys.* **105**, 151–155.

Wrigley, C.W. (1970) Protein mapping by combined gel electro-focusing and electrophoresis: application to the study of genotypic variations in wheat gliadins. *Biochem. Genet.* **4**, 509–516.

Wrigley, C.W. (1976) Isoelectric focusing/electrophoresis in gels. In 'Isoelectric Focusing'. (Ed. N. Catsimpoolas.) Academic Press, New York. pp. 93–117.

Wrigley, C.W. (1977) Characterisation and analysis of cereal products in foods by protein electrophoresis. *Food Technol. Aust.* **29**, 17–20.

Wrigley, C.W. and Shepherd, K.W. (1973) Electrofocusing of grain proteins from wheat genotypes. *Ann. N.Y. Acad. Sci.* **209**, 154–162.

Wrigley, C.W., and Shepherd, K.W. (1974) Identification of Australian wheat cultivars by laboratory procedures: examination of pure samples of grain. *Aust. J. Exptl. Agr. and Anim. Husb.* **14**, 796–804.

Wrigley, C.W., du Cros, D.L., Archer, M.J., Downie, P.G., and Roxburgh, C.M. (1980) The sulfur content of wheat-endosperm proteins and its relevance to grain quality. *Aust. J. Plant Physiol.* (in press).

11 THE DEVELOPMENT OF THE WHEAT FLOWER : GENETICS AND PHYSIOLOGY

O.H. Frankel, R. B. Knox and J.A. Considine

Introduction

The work reviewed in this paper was initiated through the discovery of a mutant in *Triticum aestivum* spp. *vulgare* which had abortive florets in specific locations in an otherwise normal inflorescence. This appeared to provide a tool for the study of the genetic control of flower morphogenesis and its role in programming the physiological processes which lead to the initiation and development of reproductive organs in plants. Such opportunities are rare indeed, due to the 'constancy' of these organs, well recognized by both Linnaeus and Darwin. However, in this instance the first discovery led to others, resulting in a genetic armoury of graduated genotypes differing by stepwise increases in organ impairment, from super-fertility to near-complete sterility. In due course the genetic system was unravelled and, for good measure, not one, but two systems, each capable of securing full fertility, and between them providing a large array of genetic and morphogenetic differentials ranging from mutants and their derivatives to various subspecies of hexaploid and tetraploid wheats.

The questions could then be asked, when the block in normal development occurred, and how gene action manifested itself? Was there a critical phase subject to elaborate genetic controls? By the application of an environmental treatment which extended the action of base-sterility genes it was possible to identify a sensitive period in the development of the spike, and subsequently a sensitive stage in the development of an individual floret. This encouraged the attempt to identify the physiological process responding to the genetic instruction. Floral differentiation is initiated by clusters of cells surrounded by cells with quite different pathways of differentiation. The accumulation of the macromolecules of development, RNA and proteins, presumably reflects changes in gene expression during the sequential events. It has been established that the target cells for flower initiation increase in RNA and protein content, enlarge and undergo cell divisions which initiate floral development. Quantitative histochemical observations have been carried out on normal and partially base-sterile apices. They confirmed the timing and established the immediate cause of the arrest of floral development, by implication indicating the function of the genetic buffering system.

This review is largely based on a series of papers, mainly of recent years, and one about to be published (Frankel *et al.*, 1969; Frankel and Roskams, 1975; Frankel, 1976; and Considine *et al.*, 1981). Genetic aspects are presented as a background to the developmental ones on which the main emphasis is placed, evolutionary aspects being largely omitted.

Genetics of floral differentiation
The Q factor

In the context of this symposium it does not seem inappropriate to recall the origins of the study which, including long intervals, has spanned most of the senior author's working life. In a plot of an English cultivar, Yeoman II, grown at the Wheat Research Institute at Lincoln, New Zealand, in 1929/30, a speltoid mutation was found in which the basal floret of most spikelets was sterile; none of the florets above the first was affected (Figure 11.1, right). Speltoid mutations, which are not uncommon in *T. aestivum* ssp. *vulgare*, got their name from

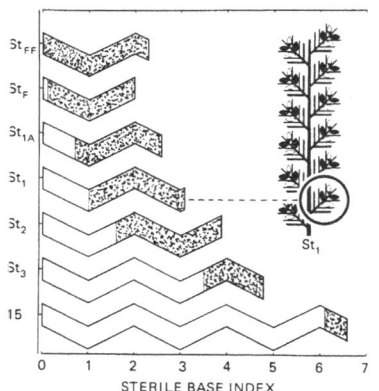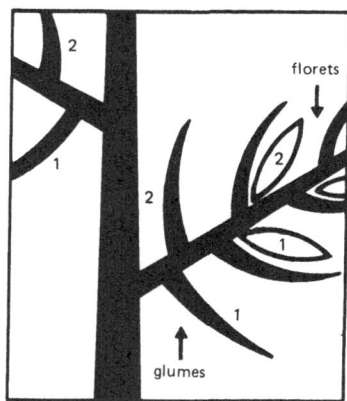

Fig. 11.1. On the right, a diagram of the original speltoid, St_1, with the lowest floret in the spikelet sterile and subsequent florets fertile. On the left an assay of speltoid stocks, from complete fertility to near complete fertility (#15, which could not be maintained). White fertile, grey sterile. From Frankel *et al.*, 1969.

their phenotypic resemblance to another hexaploid subspecies, *spelta*. Genetically, however, they are quite distinct. Speltoids result from a deletion or inactivation of the *Q* factor of *vulgare*, a complex gene or supergene responsible for the *vulgare* syndrome of soft glumes, tough rachis and short internodes in the ear. Since Yeoman has a fertile first floret, base-sterility seemed to be a function of the *Q* factor. This was confirmed in crosses with *vulgare* wheats: base-sterility was confined to the speltoid fraction in F_2 and subsequent generations of all such crosses, and was absent in homozygotes and heterozygotes for *Q*. Thus a genetic component for basal fertility is part of the *Q* complex (Frankel and Fraser, 1948).

Cryptic variation in vulgare.

A second discovery resulted from these crosses: *vulgare* varieties differ in their genetic constitution regarding base fertility. In the cross of St_1 — the original mutant — with its parent, cv. Yeoman, all F_2 speltoids were base-sterile like St_1. But in the cross St_1 x Victor, an unrelated *vulgare* cultivar, the F_2 speltoids ranged from high-level sterility of the first flower to high-level fertility. There was then an underlying cryptic variation for base-fertility in the *vulgare* varieties, which was uncovered when *Q* was deleted (Figure 11.2). From St_1 x Victor a near-fertile speltoid, St_F, was selected, which served as a high-fertility parent in many crosses. An unrelated base-sterile mutant (St_{1A}), intermediate in base sterility between St_1 and St_F, was found in a New Zealand cultivar. A common measure of basal sterility became necessary. The *sterile base index* (s.b.i.) is the mean frequency of sterile florets per spikelet in a head. It facilitates comparisons of basal sterility between genotypes, seasons or treatments. The mean s.b.i. of St_1 is 0.96 and that of St_F 0.075 (Figure 11.1, left).

2nd flower defectives

A new dimension was added when St_2, a speltoid mutant with sterile flowers in both first and second flowers (s.b.i. = 1.41) was received from Dr. A.T. Pugsley, of The Waite Research Institute. Derivatives from crosses of St_2 with St_1 and St_F (Figure 11.2) expanded the fertility-sterility range to its limits, from full fertility to high-level sterility, the extremes being represented by the stocks St_{FF} and St_3 respectively (Figure 11.1). S.b.i. for St_{FF} is 0, for St_3 3.14. Stocks other than St_{FF} and St_1, which are pressed against the barriers at 0 and 1, show a good deal of environmental variation.

Gradients

There are genetically controlled gradients which govern the pattern of sterility between the spikelets in a head. Within spikelets there also is a gradient, with a sterile zone of one or more infertile florets at the base of the spikelet axis, followed by a continuous zone of fertile florets. Exceptions to this pattern are extremely rare. A comparison of base-sterile mutants shows how, with increas-

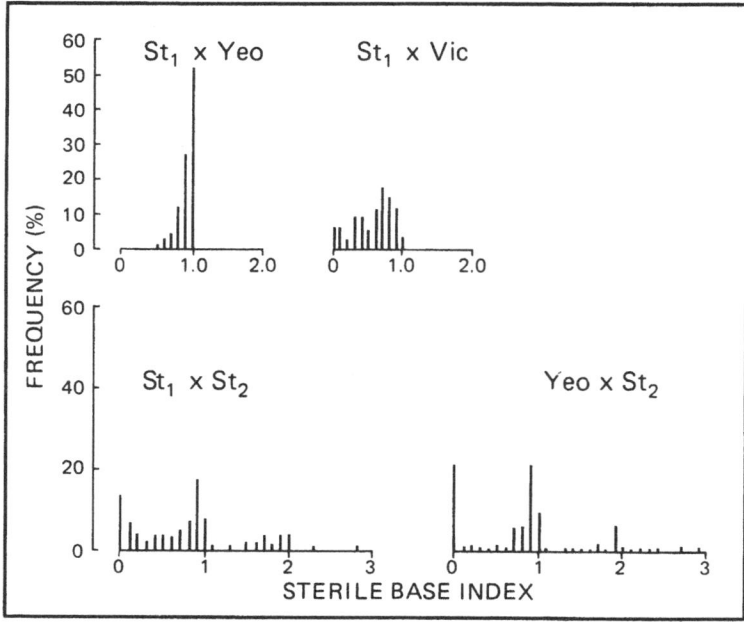

Fig. 11.2. Frequency distribution of base-sterility in F_2's: Top St_1 x Yeoman (left), and St_1 x Victor (right). Bottom St_1 x St_2 (left), and Yeoman x St_2 (right). From Frankel *et al.*, 1969.

ing sterility, the fertile zone is shifted upward into the zone of competent but not normally functional florets (Figure 11.1).

There are also temporal gradients of development. The wheat inflorescence consists of two parallel rows of lateral spikelets. Each spikelet is subtended by two normally sterile glumes. Along both sides of the axis there are bract-like lemmas which subtend the florets. The inflorescence has a temporal sequence for the development of the spikelets, the most advanced being above the middle of the primordial inflorescence. Within spikelets there is a regular gradient of development from the lowest floret upward.

A second major gene, Bs

In crosses between St_2 and first-flower-defective speltoids, St_1, St_{1A} and St_F, there is a 3:1 segregation of first and second flower defectives (Figure 11.2). No exception has been found, nor any heterogeneity between crosses. The dominant allele, *Bs* (in earlier papers called *A*), has an effect analogous to that of

Q: in the presence of Q *all* flowers are fertile, whereas in the presence of *Bs* all flowers *from the second onward* are fertile. No genes corresponding to *Bs* were found at the third flower level. *Bs* has a dosage effect on fertility, whereas one dose of Q safeguards full fertility. *Bs* thus functions like a weaker Q with respect to base-fertility. An evolutionary link is indicated by its location on chromosome 5D, a homoeologue of the Q-bearing chromosome 5A (Frankel, 1975). The *Bs* allele appears to be widespread: all *vulgare* cultivars tested carry *Bs*, and St_2 is the only known *bs* speltoid.

It is of interest to note that one *vulgare* cultivar, Chinese Spring, which is regarded as having a primitive karyotype (Riley, 1969) actually does carry a weaker Q which we have called Q'. In an F_2 with St_1, 14% of the *vulgare* fraction had some basal sterility. By selection a *vulgare* line (St_V) with a sterile base index 0.454 — half-way between St_F and St_1 — was established, and it was shown that Q'*vulgare* is subject to the same polygenic system that is responsible for the underlying variation in speltoids.

The polygenic system

All crosses involving base-sterile speltoids provide evidence of underlying variation (Figure 11.2). On the one hand this has allowed the range of sterility to be expanded beyond the St_2 level, on the other it has made possible the selection of the highly fertile *Bs* speltoid St_{FF}, as well as of *bs* speltoids with high — though not complete — *first*-flower fertility. Polygenic inheritance is confirmed by the difficulty of finding genetically stable intermediates along the sterility scale between 0 and 1.0.

Alternative pathways to basal fertility

There are then alternative pathways to complete basal fertility. One is based on Q, the other on a combination of genes among which one, *Bs*, has an effect approximating that of Q. This second system, though seemingly widespread in *vulgare,* is a cryptic and presumably relic system since it has no ostensible function in the determination of floral development, which is effectively safe-guarded by Q. Information on the origin of this second system should come from wheats that are not endowed with Q, such as the hexaploid ssp. *spelta* and the (tetraploid) subspecies of *T. turgidum* (other than *carthlicum* which has Q). In crosses of the speltoid testers with ssp. *spelta* and the tetraploid ssp. *dicoccum, turgidum, durum* and *carthlicum*, low frequencies of defectives in F_2 suggest strong buffering by the polygenic system. Corroborative evidence comes from the response to environmental 'shock' reported in the next section.

Developmental stability and destabilization

Observations of substantial fluctuations between seasons in the degree of basal sterility of some speltoid stocks suggested that the stability of floral development depended on the buffering capacity of the genetic controls. An environmental regime liable to affect floral development might provide the conditions for tests on a range of genotypes differing in the components of the genetic system. The first aim would be to ascertain the phenotypic stability of different genotypes and the direction and extent of variation which can be induced. Are there limits to the invariance of normal wheats, with or without Q? Can the zone of basal sterility be extended or reduced, i.e. can sterility be induced in normally fertile flowers, or vice versa? Should this be the case, it might prove possible to identify the stage during which the developing floret is sensitive to environmental impact, and this in turn might be a critical stage in development itself. On the basis of experience in long-day grasses (Evans, 1960), Dr. L.T. Evans suggested that the interpolation of a period of short days at moderately high temperatures subsequent to long-day induction of inflorescence initiation might be effective, and this proved to be the case.

From experiments involving varying combinations of stage of development, length of day, temperature, and duration of treatment, a standard "shock" treatment was developed which, with minor modifications, was applied to a wide range of genotypes. It consisted of a combination of a short day (8 h) and high temperature (33°C day, 28°C night), the usual duration being 15, 7 or 3 days; for details see Frankel and Roskams (1975). Results of many experiments are summarized in Figure 11.3.

(i) *Speltoids.* Basal sterility was increased in all base-sterile genotypes subjected to treatment. Means and variances of s.b.i. were significantly increased, most of all in the *bs* speltoids St_2 and St_3. Note that in St_1 sporadic 2nd flower defects occurred, which is unknown under normal conditions, and that traces of basal sterility were induced in the normally invariate, fully fertile speltoid St_{FF}.

(ii) *Non-speltoid hexaploids.* The treatment had no effect on floral development in several lines of ssp. *spelta* and *macha*. This is consistent with the low frequencies (8%) of defectives in crosses of *spelta* and St_1, already mentioned. By comparison, the fully fertile speltoid St_{FF} is marginally affected by 'shock' treatment and, crossed with St_1, has 52% defectives in F_2, showing that *spelta* possesses a more powerful buffering system than we had succeeded in giving St_{FF}.

In the tetraploid subspecies the shock treatment was as ineffective as in the hexaploids.

(iii) *Glume fertility.* Fertile florets inside the normally sterile glumes are the opposite of basal sterility, being an extension of basal fertility. This is confirm-

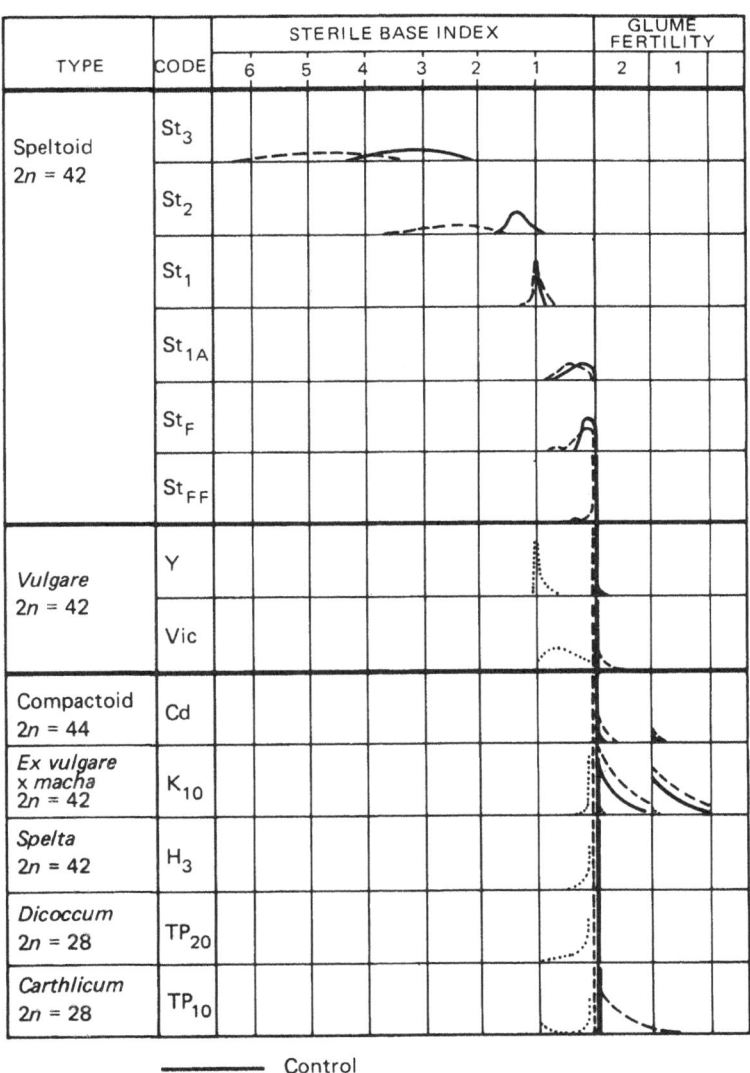

Fig. 11.3. Frequency distribution of sterile base index and glume fertility index of principal stocks used. From Frankel and Roskams, 1975.

ed by shock treatment, glume fertility being increased rather than reduced (as is basal fertility). The evidence from both environmental shock and crosses with speltoid testers indicates that floret and glume fertility are subject to the same polygenic system.

Wright (1969) reported on partially or fully developed florets in the uppermost glumes of many *vulgare* cultivars. He had also found a compactoid mutant (Wright, 1958), which we named *Cd*, with a higher frequency of fertile glumes, which is not surprising since it is tetrasomic for the 5A chromosome, hence for *Q*. We found an even higher level of glume fertility in *vulgare*-like derivatives (with 42 chromosomes) from a cross *vulgare* x *macha*, kindly supplied by Professor H. Kuckuck. The most glume-fertile of these, K_{10}, is included in Figure 11.3.

The figure shows an increase in glume fertility in all *vulgare* and *vulgare*-derived stocks, corresponding to their polygenic complement as ascertained in crosses with St_1. Thus the effect on Victor is greater than on Yeoman which, as we have seen, has a polygenic complement corresponding to St_1 (see Figure 11.2), and greater still on K_{10} which has only traces of basal sterility (and glume fertility) in its St_1 cross. Here the parental polygenic systems acting in opposite directions virtually cancel each other out, resulting in a phenotypic balance approaching that of the normal wheats. This is a clear demonstration of a common genetic control of the normal floral fertility and glume sterility in common wheat.

The induced glume fertility in *carthlicum* provides yet another sidelight, revealing the effect of *Q*, superimposed upon the polygenic background of the tetraploids.

In summary, when the environmental shock treatment is applied to the experimental tester stocks, it may drastically increase either basal sterility, or glume fertility, depending on the genetic composition. When applied to genotypes with a long evolutionary history - including spp. *vulgare, spelta* and *dicoccum*, it has, at most, a marginal effect, the only exception being *carthlicum*. Here, it seems, the combination of *Q* with the normal tetraploid polygenic complement results in an excessive level of buffering which is expressed in the high level of glume fertility induced by the treatment.

The sensitive stage of floral development

Having found effective environmental conditions for inducing defects in florets which otherwise would be normal, it should be possible to define the temporal conditions under which this transformation occurs, i.e. to identify the stage of development which is sensitive to the treatment. We have already recognized gradients for basal sterility within the ear and within spikelets. There

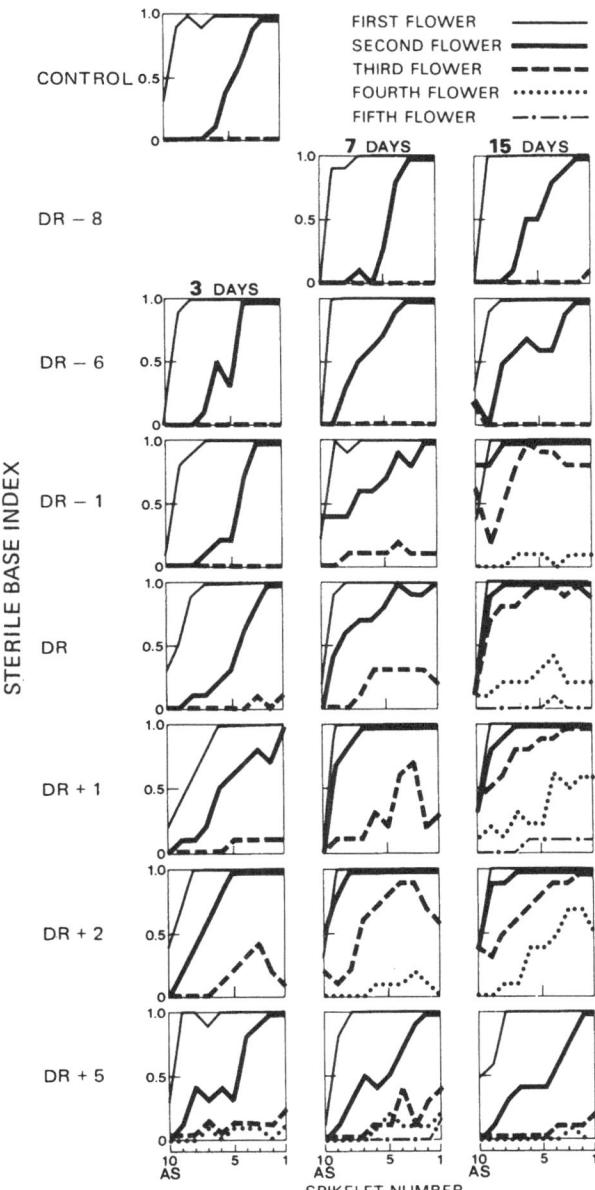

Fig. 11.4. 'Shock' treatments of St_2. Effects of treatments commencing at consecutive stages (columns) before, during and after spikelet initiation ('double ridge stage', d.r.), from d.r.-8 to d.r. + 5 days. Graphs illustrate sterility gradients of consecutive flowers, from the basal spikelet (1) to the apical spikelet (a.s.). From Frankel, 1976.

is also a developmental gradient within a wheat inflorescence. By varying the inception and/or duration of the treatment, it should be possible to identify the sensitive period for the ear as a whole, and also for individual spikelets or florets, in accordance with the temporal pattern of their development. If the sensitive period is found to coincide with floret initiation there would be presumptive evidence that floral competence is determined at this stage.

The experiments were designed with two basic variables, inception and duration of treatment. Earlier observations had shown that treatments starting at or about the initiation of the most advanced spikelets were the most effective. Accordingly, treatments started at from 8 days before to 10 days after spikelet initiation, with durations varying in different experiments. The effects of treatments were identified by dissection after ear emergence. The stock principally used was St_2, with the first flower defective in all but the top spikelets, the second flower with a gradient from full sterility at the base to full fertility at the top, and the third flower fully fertile (top of Figure 11.4, control).

In the experiment illustrated in Figure 11.4, treatments started successively from 8 days before till 5 days after spikelet initiation (columns), with durations of 3, 7 and 15 days (rows). Development proceeded during treatment, as is evident from the effect increasing with duration. The results show, first, that there is a *sensitive period* for the spike as a whole, from < 6 days before to 6 or 7 days after spikelet initiation. Second, they show that successive florets, from the second to the fourth (and, marginally, the fifth), and successive spikelets (from the bottom towards the apex of the spike) reach a *sensitive stage* in a well defined temporal sequence, and that prior and subsequent to this stage the treatment has no effect.

So far we have classified florets either as normal or as sterile (or defective). However, a more detailed classification of floret defects is required if defects identified at ear emergence are to be traced back to the differentiation of the floret. Hence the nature of the defects and their positioning within the spike need to be specified as closely as possible.

Four grades account for the great majority of florets at ear emergence, intermediates being rare (Barnard, 1955b).

1. perfect
2. one anther missing
3. flower organs malformed, fused or twisted
4. empty

Grades 1 and 2 are fertile, 3 and 4 sterile.

Figure 11.5 illustrates the floral profiles of St_2 (rows 1 and 2) and St_3 (rows 3 and 4), control in the upper and shock-treated in the lower row. Wholly sterile flowers are omitted. In a transition zone between the empty and fertile

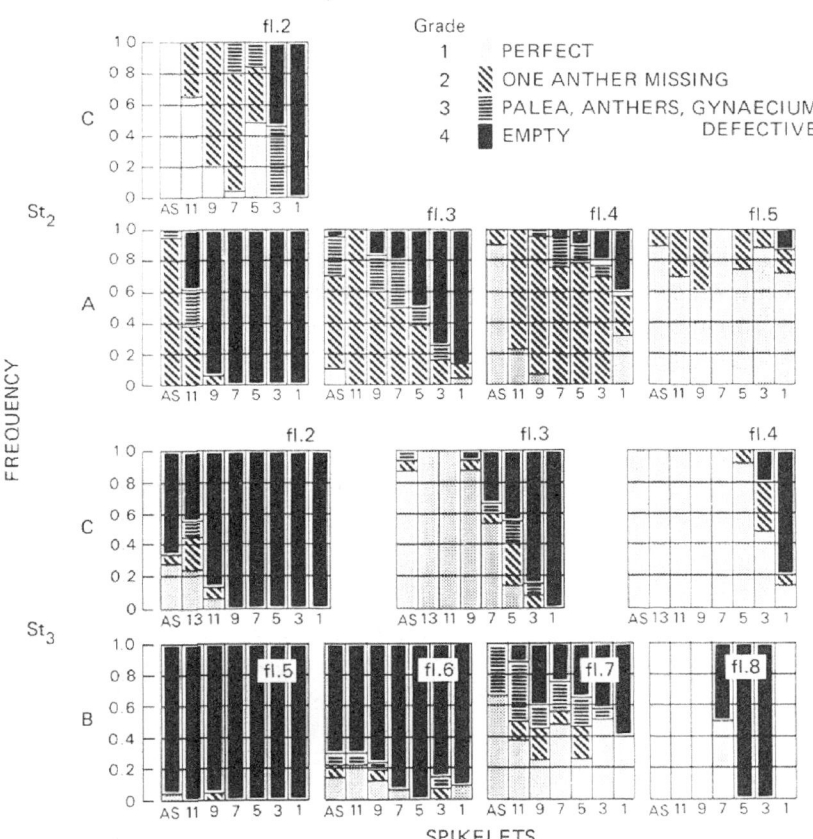

Fig. 11.5. Experiment with St_2 (rows 1 and 2) and St_3 (rows 3 and 4); rows 1 and 3 control, rows 2 and 4 'shock' treatment. Sterility-fertility grades in successive spikelets, from 1 (bottom) to a.s. (apical spikelet). Legend for grades in top right corner. Wholly sterile flowers are omitted (e.g. flower 1 in St_2). C control, A and B treatment. From Frankel, 1976.

zones we find defective florets in a regular sequence from 4 (empty) to 3, 2 and 1 (perfect), the defective zone being more extensive in St_2 than in St_3. On the basis of this information it is possible to indicate the sites of defective florets and to follow their initiation and further development. An example is illustrated in Figure 11.7 (St_2 control). The second floret is clearly smaller and lower in

RNA content than is the next younger one, and is interpreted as being grade 3. At a slightly more advanced stage such florets are readily identified as being backward in development in relation to younger florets in the same spikelet (see below).

Floral Initiation and the sensitive stage

The experiments on the effects of sequential shock treatments during floral development had shown that treatments were effective only during a limited period, which was reached by individual florets in accordance with the temporal gradients within the ear. It was apparent that the sensitive period, as indicated by treatment response, was associated with a specific stage in the differentiation of individual florets.

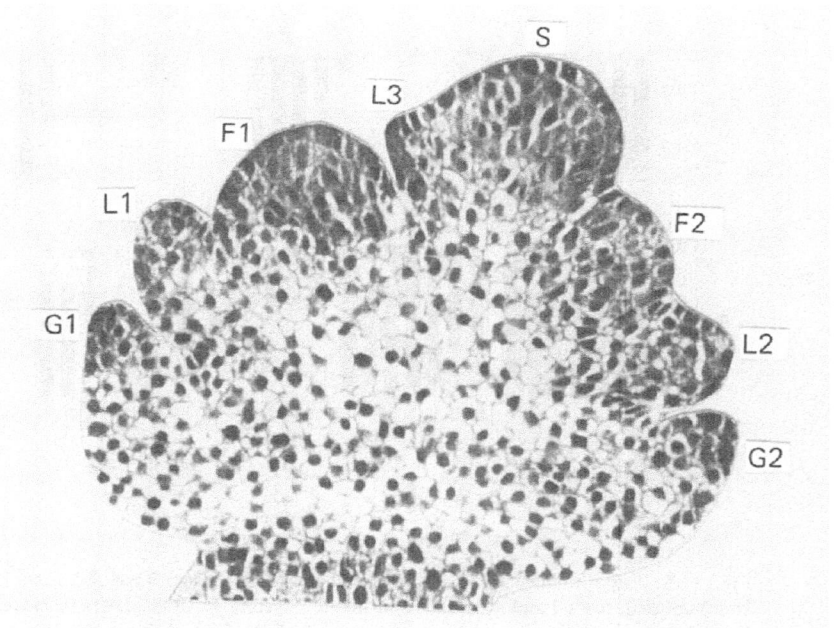

Fig. 11.6. Gabo spikelet. G glume, L lemma primordium, F flower primordium, S summit.

Floret differentiation is initiated by the differentiation of the subtending bract, the lemma, which is followed by the initiation of the floret itself and of the next higher lemma in the spikelet (Figure 11.6). The developmental rhythm of the initiation and growth of successive lemmas provides an objective scale for the floral development of normal, defective or totally aborted florets. These three classes are readily distinguished at an early stage: abortive florets through the total absence of periclinal cell divisions, and defective florets by their reduced growth rate by comparison with a distal normal floret (Figure 11.7).

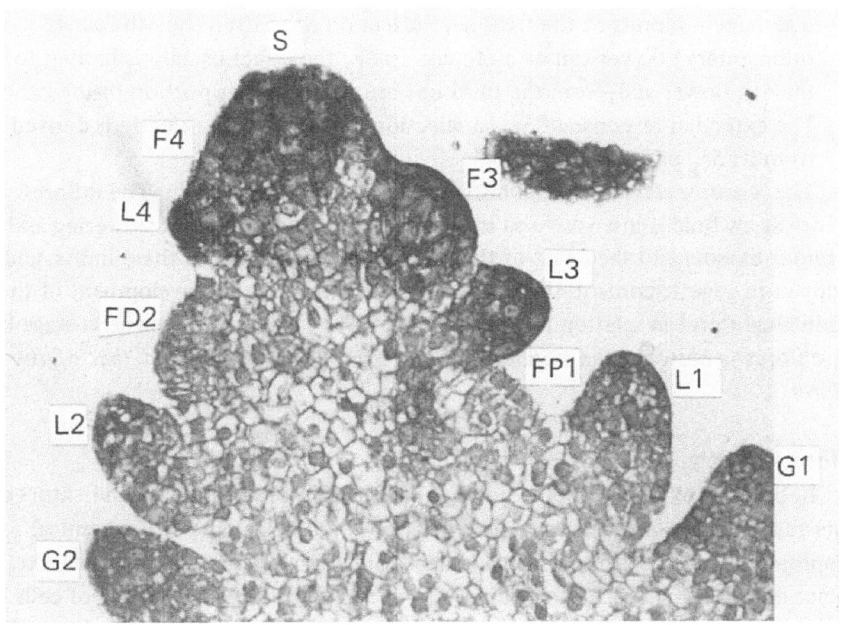

Fig. 11.7. *St$_2$* spikelet (untreated). G glume, L lemma primordium, F flower primordium, FP flower position (empty), FD defective flower primordium, S summit.

Evidence on initiation and differentiation comes from apices periodically sampled from experiments prior to, during and subsequent to the application of treatments, commencing from the earliest stage of floral development — the formation of lemma initials — and continuing until the fertility status of the youngest florets could be identified or the end of treatment, whichever was the later. The main findings were:

1. Each flower is sensitive to treatment for a limited period (the *sensitive period*). Treatment effects are confined to the treatment period, with only a marginal after-effect (no more than one day).
2. There is a *sensitive stage* for each individual floret, starting with the differentiation of lemma initials and ending with the stage at which in normal

flowers cell enlargement initiates floret differentiation. It appears to be less than 24 hours.

3. The differences in the differentiation of normal, defective and empty florets make identification possible within 2 or 3 days of floret initiation. It corresponds closely with frequencies at maturity.

4. There are no limits to the treatment effect on St_3. Even the 8th (partly rudimentary) flower can be affected. In St_2 the effect usually is limited to the 4th flower and, from the third onward, to the lower portion of the ear. The extended response of St_3, a selection from St_2 x St_1, clearly is derived from its St_1 parent.

The *sensitive stage* for floral initiation can now be identified. The inflorescence as a whole is insensitive to treatment prior to induction of flowering and again subsequent to the onset of floral differentiation. Within these limits, and subject to genetic control, the decisive factor is the stage of development of the individual floret in relation to the incidence of treatment. The sensitive stage of the floret is *between the initiation of the subtending lemma and that of the floret.*

Histochemistry

In order to exploit fully the potential of the speltoid mutants as indicators of the genetic control of flower differentiation, cytological and histochemical approaches have been used. This is necessary because the changes that take place during floral evocation at the shoot apex are localized in clusters of cells in the axils of the lemma primordia, and not in the entire spikelet primordium. In this study, changes in nucleic acid content of these cells have been followed using quantitative histochemical methods.

Qualitative histochemistry

At the base of each spikelet is a pair of sterile glume-primordia (Figure 11.6), followed by the differentiating florets arranged alternately along the axis. The lemma primordia subtend each floral apex and the cells stain an intense purple colour with azure B, indicating a high RNA content (Figure 11.8). The floret primordia differentiate by cell division into a hemisphere of densely-staining cells. In the primordium differentiating at the spikelet apex, the cells are always subtended by a lemma primordium, and are evident by the swelling of the hypodermal cells and their intense purple azure B staining (Figure 11.8). Floret development is rapid, the basal floret (1) often having differentiated floral organs while floret 2 is still only a hemisphere of cells and floret 3 is not apparent (Figure 11.6). A feature of the RNA staining by azure B is that the purple staining is found not only in the cytoplasm of the meristematic cells but

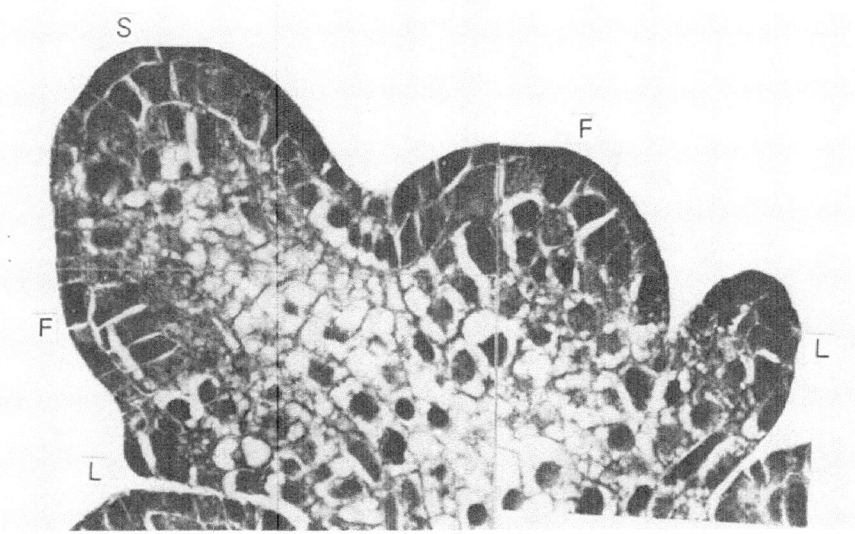

Fig. 11.8. Gabo spikelet, detail, symbols as in Figure 11.6.

also abundantly in the nucleus where it is present in nucleoli, and around the chromatin. In St_3 mutants sterility sets in at the earliest stage in floral init-iation. Lemma primordia develop normally but the hypodermal cells of the potential floret primordia fail to undergo the periclinal cell divisions necessary to initiate a new floral primordium, and show only faint staining of the nuclei with azure B, indicating a low RNA content (Figure 11.9). Upper (fertile) florets differentiate normally as in Gabo. St_3 spikelets showed at least the first two floret primordia without normal development, and the contrast between these and the upper fertile primordia is striking (Figure 11.9). The extreme of basal sterility is reached in treated St_3 apices in which all 6-7 primordia that had developed are sterile (Figure 11.10).

In these longitudinal sections, it is apparent that there is both a reduced RNA content in base-sterile florets, and a reduction in cell number in the floret primordia which remain as discs, failing to divide and expand to form the dome characteristic of fertile florets.

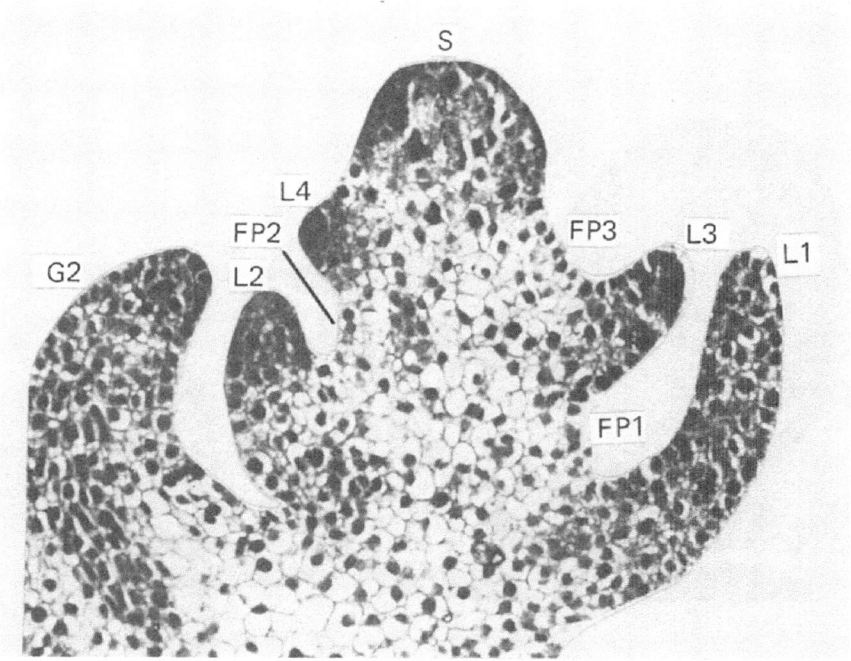

Fig. 11.9. St_3 young spikelet. Symbols as in Figure 11.7.

Quantitative histochemistry

A stereological analysis has been carried out to demonstrate proliferation of epidermal and hypodermal cells in Gabo and St_3 in both floral and subtending lemma primordia. The data are presented as the estimated number of epidermal cells over the floret dome (Gabo) or disc (St_3) plotted against the number of cells intercepted in the plane of the section as an index of floret development. The graph shows that proliferation of epidermal cells is severely restricted in St_3 whereas those of Gabo proliferate exponentially (Figure 11.11). An interesting feature of these data is that the floret primordia epidermal cell numbers for both genotypes have an intercept close to one, suggesting that the differences are initiated at the earliest stage of development.

Quantitative comparisons of total nucleic acid content support the visual appraisal of RNA content in sterile and fertile floret primordia. The scans of absorbance of individual spikelet primordia reveal intense absorbance in the meristematic regions: the lemmas, the apical dome and the dome of presumptive fertile florets (Figures 11.12, 11.13). The intensity of absorbance in the

Fig. 11.10. St_3 (treated), older spikelet. Five empty florets, 6th floret different-iating.

regions of lemma primordia is similar for lemmas in both presumptive fertile and presumptive sterile primordia, but the absorbance in the regions of the florets is strikingly different in the two types.

Close inspection of the scans discloses small areas of intense absorbance in the region of the apical domes of Gabo spikelets which are less evident in St_3, suggesting that differences in RNA content may appear at the earliest stages of differentiation of the primordia. This contention is supported by a regression analysis of the total RNA content of recognisable floret primordia with the peripheral cell number of the subtending lemma as an index of time or stage of development of the primordium (Figure 11.14). The regression lines fitted demonstrate that the zero intercept values of nucleic acid content are higher in the fertile Gabo spikelets than in the basal sterile St_3 spikelets right from the time of their initiation. Also, as suggested by the individual scan data (Figures 11.12, 11.13) the rate of accumulation of nucleic acids is limited in St_3 and the rate constant is about one-tenth to one-twentieth of that of cv. Gabo.

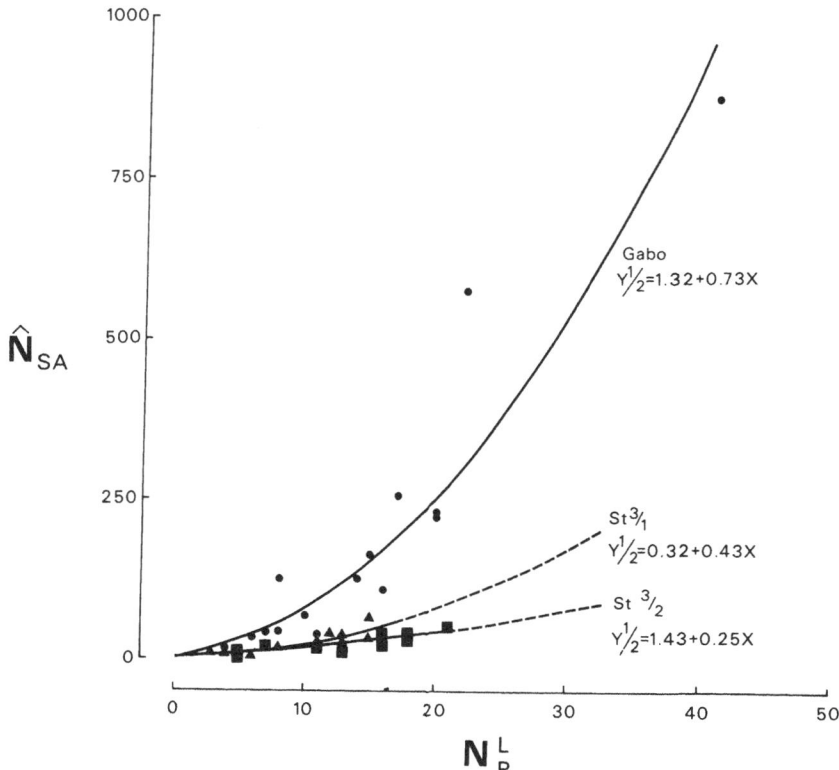

Fig. 11.11. Estimated total epidermal cell number (N_{SA}) of one floret primordium of Gabo and two of St_3, plotted as a function of the number of epidermal
cells transected in the perimeter of the subtending lemma primordium (N_P^L). The
value N_{SA} was calculated from the number of epidermal cells transected in the
perimeter of the floret primordium assuming that the primordium was in the form
of a hemisphere. The curves were analysed by linear regression by taking the
square root of the N_{SA} and thus converting it to a coded linear value. The
regression equations were assessed by analysis of covariance. The regression line
for Gabo differed significantly from those for the St_3 primordia at $P < 0.01$.
Least significant difference for the regression coefficient 'b' was 0.085 at $P = 0.05$.

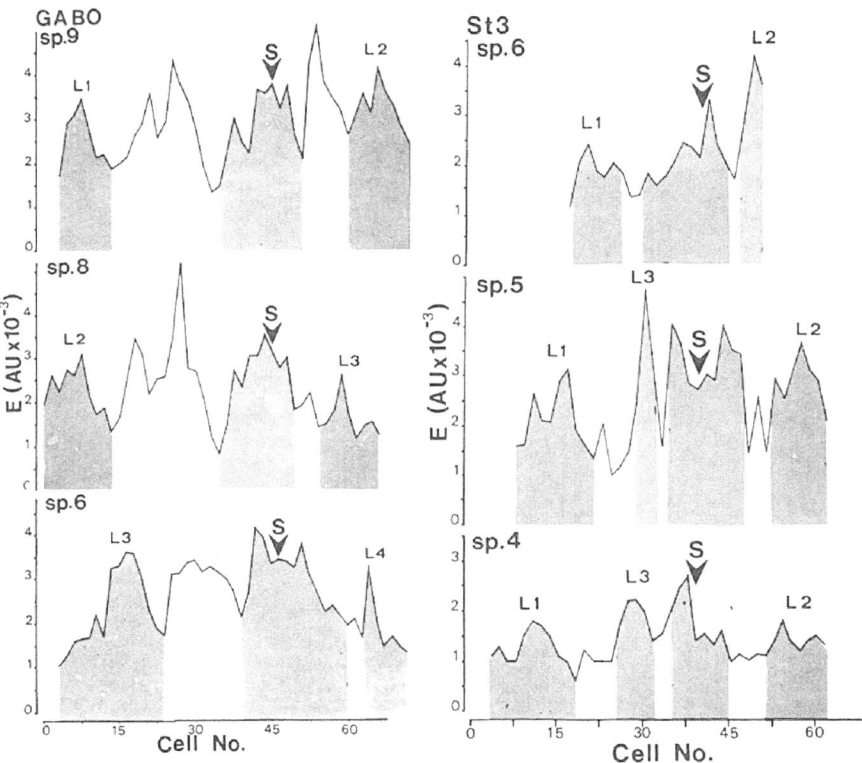

Fig. 11.12. Microspectrophotometric data from individual scans of epidermal and hypodermal layers of Gabo showing absorbances (in arbitrary units) for total nucleic acids in the various lemma and floret primordia. Microspectrophotometry was carried out by measuring the absorbance of 50 μ m diameter circular overlapping layers. In this way a continuous record of the RNA content of the surface two layers of cells was obtained. The position of the summit tip is indicated by an arrow. EAU = absorbance in arbitrary units. For explanation of cytoscan measurements refer to Considine, Knox and Frankel, 1981.

Fig. 11.13. (Right hand side). As for Figure 11.12; data for St_3.

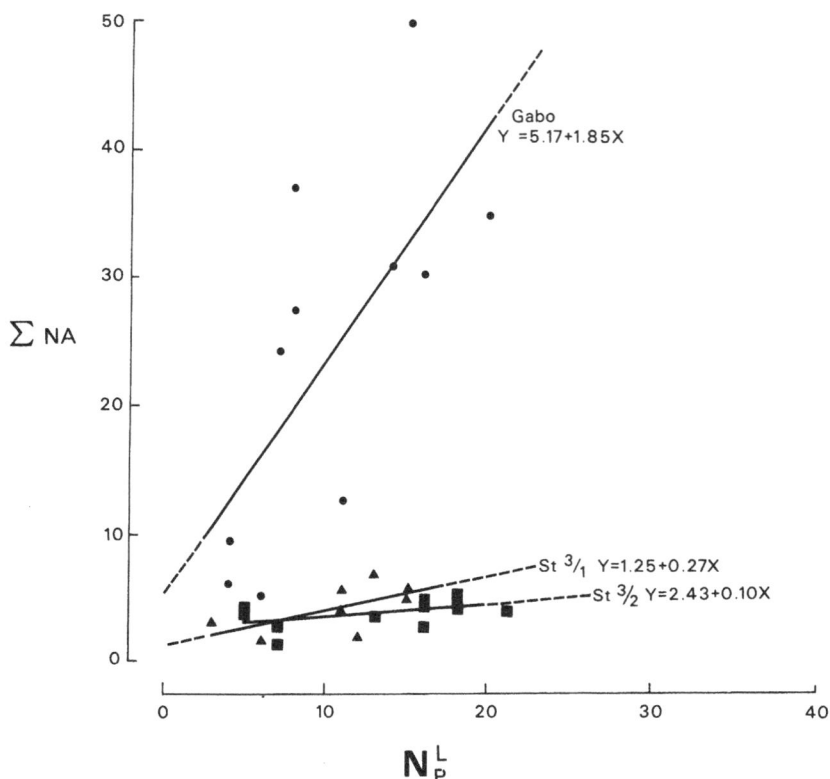

Fig. 11.14. Total nucleic acid content (ΣNA, arbitrary units) measured in florets of Gabo and St_3 spikelets plotted as a function of the number of epidermal cells transected in the perimeter of the subtending lemma primordium (N_P^L). Two spikelets of St_3 were measured. The regression equations were computed from the data and used to plot the lines of best fit. Analysis of covariance demonstrated that the regression fitted for the Gabo spikelet was statistically different from the two St_3 spikelets at $P < 0.01$. Least significant difference for the regression 'b' was 0.49 at $P = 0.05$.

These data demonstrate that in the completely sterile floret primordia of St_3, reduced development is not associated only with a reduced rate of accumulation of RNA, but is also apparent as a reduced rate of cell proliferation. These data show that the expression of the genes for basal sterility involves multiple inter-related effects on the floret primordia:

(1) cessation of RNA accumulation
(2) restriction of cell proliferation by anticlinal cell divisions
(3) elimination of periclinal cell divisions in the hypodermal cells.

Systematic scans of floral positions likely to result in defective florets of grades 2 and 3 (Figure 11.5) have yet to be made. They may yield further quantitative information on the role of RNA in relation to floral differentiation.

Conclusions

The speltoid series of mutations has provided a unique genetic system for investigation of the control of flower development. For the first time, specific mutations affecting the processes governing flower formation in certain defined parts of the inflorescence have been available for cytological and histochemical studies. These have revealed temporal and spatial changes in the expression of gene products important for growth and development in both fertile and potent-ially sterile florets right from their inception. We have found no evidence for a subhypodermal origin of the florets as proposed by Barnard (1955a).

The genetic system involves the major locus Q in regulation of floret fertility. When Q is absent, the Bs and bs alleles, in association with a polygenic system, express the fertility level in the speltoid phenotypes. This genetic system interacts with the environment, adverse photoperiodic and temperature regimes enhancing the expression of base sterility in each spikelet. These physiological experiments have also revealed the existence of a sensitive period in development of the inflorescence apex — at or soon after spikelet initiation — when floret sterility may be induced by exposure to particular environmental conditions. The floret primordia also have a sensitive stage when they are susceptible to environmental modification. An analogous situation has been reported by Heslop-Harrison (1961) during inflorescence development in maize, *Zea mays*, where the presumptive male flowers of the tassel are switched to female flowers by adverse photoperiod and temperature treatments. There is, therefore, evid-ence for the existence in both these systems for a period during flower develop-ment when the primordium is uncommitted to a particular developmental pathway.

A feature of the present data is that the sterility genes are expressed at the very earliest stages of floret morphogenesis - at about the time of initiation of the precursor cells at the apical dome of the spikelet, well before formation of a recognisable floret primordium. The detection of changes so early in different-iation, was made possible by the use of a combination of quantitative histo-chemical and stereological techniques, by which changes in cellular RNA could be inferred at stages before they were detectable visually. Quantitative differences in RNA levels have been reported for other floral systems (Lyndon, 1970; Cecich and Horner, 1977; Varnell and Vasil, 1978a and b) but none has been able to demonstrate effects so early in differentiation. Indeed, stereologic-al analysis has made it possible to investigate the specific time during develop-ment when initiation of the floret primordium actually occurs. A specific, nascent tissue - the dome of the spikelet primordium - is involved, and the floret primordium appears to be produced from as few as perhaps one or two cells lying between the lemma primordial cells and the apical dome of the spikelet.

An important feature of these data is the observation that lemma initiation always preceded floret formation, and is an essential step even for sterile florets. This observation is analogous to the finding in the inflorescence primordium of *Lolium temulentum* that bract initiation always precedes spikelet primordium formation in its axils (Knox and Evans, 1966). It is also noteworthy that initiation of a fertile floret in a spikelet always leads to subsequent florets being also fertile, evidence in favour of Waddington's (1962) concepts of the canaliz-ation of development. Fertile floret formation requires that cells of this tissue maintain the meristematic characteristics of the apical meristem as they are displaced from the tip of the apical dome, i.e. RNA synthesis is maintained at a high level and the cells maintain a high rate of anticlinal cell division. Fertile floret differentiation then follows and involves stimulation of expansion of hypodermal cells and initiation of the formation of a new growth axis through periclinal cell divisions in the hypodermal cells.

It is interesting that the sterility genes and adverse photoperiod and temper-ature treatments do not induce abortion of entire spikelets or even of lemmas, but act to shift the site of formation of the first fertile floret to a more distal lemma axil. Taken together, one explanation for these observations is that a gradient of diffusible inhibitory or blocking substance moves acropetally from the body of the plant and that the apical meristem of the spikelet acts as the source of a promotive substance or hormone, producing a bidirectional gradient of promotive and inhibitory substances. This model is an expansion of that proposed by Fraser (1950). The normally sterile but — as we have seen — po-tentially fertile glumes at the base of the spikelet could be seen as further evidence for the shifting balance between the controlling substances. The evidence also suggests that a fertile floret modifies the gradient by either block-ing diffusion of the inhibitory substances or by supplementing the apex as a source of stimulant. Candidates which readily come to mind are cytokinins produced at apical meristems, and can act to induce meristem formation and abscisic and phaseic acids which increase in vegetative tissues in response to environmental stress and short days.

The specific nature of the response of the floret primordial cells is also conceivable in terms of the target cell concept. These studies provide a basis for elucidating the mechanisms which control the first steps in organ formation, the formation of particular target cells at the apical meristem and their response to a stimulus to either proceed along the complex path to floret formation or to abort. Whether the genes act to modify the responsiveness of these target cells or modify the basal level of stimulus is now open to investigation, as is the identity of the messenger molecules.

References

Barnard, C. (1955a) Histogenesis of the inflorescence and flower of *Triticum aestivum* L. *Aust. J. Bot.* 3, 1–20.

Barnard, C. (1955b) Sterile base florets in *Triticum*. *Aust. J. Bot.* 3, 149–164.

Cecich, R.A. and Horner, H.T. (1977) An ultrastructural and microspectro-photometric study of the shoot apex during the initiation of the first leaf in germinating *Pinus banksiana*. *Am. J. Bot.* 64, 207–222.

Considine, J.A., Knox, R.B. and Frankel, O.H. (1981) Quantitative histo-chemical changes in cellular RNA associated with basal sterility during floral development in wheat, *Triticum aestivum*. In preparation.

Evans, L.T. (1960) The influence of environmental conditions on inflorescence development in some long-day grasses. *New Phytol.* 59, 163–174.

Frankel, O.H. (1975) Base-sterile speltoids: the location of the *Bs* gene of *Triticum aestivum*. *Proc. R. Soc. Lond. B Biol. Sci.* 188, 163–166.

Frankel, O.H. (1976) Floral initiation in wheat. *Proc. R. Soc. Lond. B Biol. Sci.* 192, 273–298.

Frankel, O.H., and Fraser, A.S. (1948) Basal sterile mutants in speltoid wheat. *Heredity* 24, 391–397.

Frankel, O.H., and Roskams, M. (1975) Stability of floral differentiation in *Triticum*. *Proc. R. Soc. Lond. B Biol. Sci.* 188, 139–162.

Frankel, O.H., Shineberg, B., and Munday, M. (1969) The genetic basis of an invariant character in wheat. *Heredity* 24, 571–591.

Fraser, A.S. (1950) Basal sterility in wheat. *Nature (Lond.)* 165, 653.

Heslop-Harrison, J. (1961) The experimental control of sexuality and inflores-cence structure in *Zea mays* L. *Proc. Linn. Soc. Lond.* 172, 108–123.

Knox, R.B., and Evans, L.T. (1966) Inflorescence initiation in *Lolium temulentum* L. VIII. Histochemical changes at the shoot apex during induct-ion. *Aust. J. Biol. Sci.* 19, 233–245.

Lyndon, R.F. (1970) DNA, RNA and protein in the pea shoot apex in relation to leaf initiation. *J. Exp. Bot.* **21**, 286–291.

Riley, R. (1969) Evidence from phylogenetic relationships of the types of bread wheat first cultivated. In 'The Domestication and Exploitation of Plants and Animals'. (Eds. P.J. Ucko and G.W. Dimbleby.) Duckworth & Co., London. pp. 173–176.

Varnell, R.J., and Vasil, I.K. (1978a) Experimental studies of the shoot apical meristem of seed plants. 1 Morphological and cytochemical effects of IAA applied to the exposed meristem of *Lupinus albus. Am. J. Bot.* **65**, 40–66.

Varnell, R.J., and Vasil, I.K. (1978b) Experimental studies of the shoot apical meristem of seed plants II. Morphological and cytochemical effects of kinetin applied to the exposed meristem of *Pinus elliottii. Am. J. Bot.* **65**, 47–49.

Waddington, C.H. (1962) 'New Patterns in Genetics and Development.' Colombia University Press, New York.

Wright, G.M. (1958) Grain in the glume of wheat. *Nature (Lond.)* **181**, 1812–1813.

Wright, G.M. (1969) A developmentally unstable character in wheat: glume fertility. *N.Z. J. Bot.* **7**, 30–35.

12 THE INTERACTION BETWEEN THE PHYSIOLOGY AND THE BREEDING OF WHEAT

J.B. Passioura

The breeding of wheat is a very practical affair. Its success or failure is judged, ultimately, in the exacting environment of the farmers' fields. Physiology is not so practical, and its success is judged primarily on how enlightening it is, and secondarily on whether this enlightenment provides leads, or perhaps tools, for the scientific practitioners of wheat-growing, the breeders and the agronomists, to use or to follow up.

There is no doubt that physiological understanding provides a major part of the framework within which the breeder develops his generally implicit thoughts about what are ideal wheat plants. There is, however, much doubt about whether or not physiology has had, or ever can have, a major direct influence on plant breeding. Overt attempts by physiologists and breeders to collaborate have led, almost without exception, to failures. The reasons for these failures have become clear with hindsight for each particular case. I do not intend to discuss these attempts in a systematic manner. What I do want to do is to draw some general conclusions about why they have failed and to see if there are any general lessons to be learned about how we might improve the interaction between the breeding and the physiology of wheat. My remarks will be somewhat philosophical, and may, on that account appear unworldly, but I nevertheless believe that they have considerable implications for the management of biological research. I think it appropriate to discuss the subject in this way, because this meeting is being held in honour of Otto Frankel, and because he has had an enduring interest in both the philosophy and the management of science. I do not expect him to agree with what I have to say, but I hope he will enjoy disagreeing with it.

The enclosed gas as an archetypal example of an organised system

As my seemingly irrelevant starting point, I take the behaviour of enclosed gases. The relevance will, I hope, emerge. The behaviour can be discussed in two distinct ways: (1) in terms of the gas laws, which treat the gas as a whole, and tell us that the product of the pressure and the volume of a fixed amount of gas is directly proportional to its temperature, and (2) in terms of the properties of the constituent molecules of the gas. The enclosed gas provides us with the only near-perfect example of the way our minds apprehend complex systems by analysing them, and synthesising them, into and from, what we perceive to be their components. The excellence of this example arises from the excellence of the theory that enables us to explain the behaviour of a gas from the behaviour of its components. This theory is statistical mechanics.

I would like to dwell on this simple example for a while, and use it to illustrate some of the properties of hierarchically organised systems, before returning to that highly organised and highly complex system, the wheat plant. When we analyse an organised system, we in effect divide it into conceptual layers. We create concepts that are typically peculiar to each layer, and may have meaning only in one layer. The ordinary concept of pressure, for example, has no meaning if we are studying the behaviour of individual molecules. When we say that we understand the behaviour of such a system we usually mean that we have resolved it into components, and have discovered interactions between these components which, when suitably integrated, reproduce the behaviour we are trying to explain. Statistical mechanics is the tool that can integrate the behaviour of molecules to produce the gas laws. But there are many ways in which integrations of this sort can be carried out, and we therefore have to specify certain constraints before we can carry out the appropriate integration. In the case of our gas there are three major constraints. The first is that there is a fixed amount of gas; the second is that the gas is entirely enclosed within a container (the gas laws do not apply to the earth's atmosphere, but statistical mechanics does); and the third is that of the many different properties of the molecules that we might consider, we restrict our interest to their kinetic energies of translation, i.e., to the energies associated with their movement through space. As Rosen (1969), from whom I have borrowed this example, has pointed out, there seems to be no way of deciding what these constraints should be on the basis of molecular information alone. We must discover the gas laws, by observing the behaviour of a gas, before we can set about explaining them. If plant physiology is indeed a retrospective science, as plant breeders have been alleged to remark (Evans, 1977) then so is statistical mechanics, at least as far as the gas laws and several other classical physico-chemical laws (e.g. Stefan's, Dulong's and Petit's) are concerned.

If, as Rosen (1969) persuasively argues, phenomenological discoveries in hierarchically organised systems can virtually only be made at the level at which they apply, it does not at all mean that it is pointless to reflect on behaviour at adjacent levels in the hierarchy. I have argued at length elsewhere (Passioura, 1979), that such reflection is very important in guiding and stimulating the imagination when we are faced with systems whose organisation is as complex as that of a plant.

In considering the behaviour of a gas we have thus far been concerned with only two levels of organization. If we were interested in very fine detail we might start considering a third level, say the internal behaviour of the molecules as manifest in the relative movements of their constituent atoms. It is very unlikely that we would concern ourselves with a fourth layer, say, the internal behaviour of the atoms — that is a task of a nuclear physicist, not a statistical mechanic. But when we are dealing with a biological system, there are at least half a dozen levels that we might have to consider. Most physiologists who have worked with wheat have given serious thought to its behaviour across most of the complete spectrum of levels — from the community of plants that is a wheat crop, through the descending levels of whole plant, organ, tissue, cell, membrane, and molecule. I think it is this very versatility that has led to many of the barren marriages between physiology and breeding. Breeding is concerned with improving the performance at the community level. The applied physiologist has a similar concern, even though his work is typically not being carried out at the community level, but at one of the many levels below. He is therefore tempted to leap in one bound from whatever level he happens to be working on, to the behaviour of the crop in the field, and will try to discover traits, at his level, that he thinks will improve performance in the field. In so doing, his behaviour is analogous to that of a physicist who is trying to discover the gas laws by riding around on the back of a molecule. In fact his behaviour may be even more absurd than that; it may be analogous to that of a physicist searching for the gas laws while riding around on the back of a proton.

It is here, then, that I believe the infertility of the marriage between physiologist and breeder lies — that the physiologist has been so entranced with the marvellously intricate processes that occur within plants that he has made his initial dissection of the crop plant too fine. He has typically been using a scalpel where he should have been using secateurs or perhaps even an axe. I would like now to pursue this theme by considering some specific examples. My examples will be concerned with the performance of water-limited wheat plants, and there are two reasons for this. The first is that this is the area of my own research, so that I have given more thought to it than to other types of limitations to growth. The second is that water-stress is a somewhat simplifying circumstance and enables us to ignore some of the subtle but nevertheless important facets of behaviour of the well-watered wheat plant.

Some examples of infertile interaction
The size of the root system

Until quite recently it was almost universally assumed that a large root system would improve the drought resistance of a plant. The account in Kramer

(1969) is an excellent example of this view, which, in terms of our previous discussion on levels of organization, implies that we are predicting the behaviour of a crop from a knowledge of the behaviour of a whole plant or perhaps an organ. A well-known Canadian wheat-breeder thoroughly espoused this view and indeed set up some controlled-environment experiments in which he showed it to be true (Hurd, 1974). But a few moments reflection on the interaction between the root and the shoot, should convince us that there is no good *a priori* reason why this view should be universally true.

There are two major assumptions embedded in this view. The first is that the water supply is large, but not easily accessible. The second is that the expenditure of hard-won assimilate in the roots does not represent a major diversion from the grain. The first assumption is demonstrably wrong in many wheat-growing areas, where the available water-supply is generally restricted to the top metre or so of the soil, and during a drought is completely depleted by the crop. The second assumption must also be often wrong, for although the roots typically contain only about 10% of the total dry matter of a well-watered wheat crop at harvest, they may contain a much greater proportion than this in a droughted crop, and furthermore the proportion of the plant's assimilate that they expend may be much more than would appear from the root:shoot ratio, for there is evidence that roots respire much more rapidly (per unit dry weight) than does the shoot (Sauerbeck and Johnen 1976). Thus we could argue that, contrary to the doctrine in the previous paragraph, the performance, i.e., the grain yield, of a droughted wheat crop could conceivably be improved by reducing the size of the root system. Hurd's (1974) view is undoubtedly right for the Canadian wheat-belt where deep soils contain much water beyond the reach of the root system of the spring wheat plant, but that does not mean that we can extrapolate to other drought-prone environments.

Hurd's controlled-environment experiments, incidentally, provide a good example of how easy it is for a physiologist to set up a self-fulfilling prophecy. His comparison of genotypes having a wide range in vigour of the root system was done by growing single plants in huge pots containing many times as much soil as a single plant would have untrammelled access to in the field. Considering that the pots were not watered after sowing, it would seem certain that those plants with the most vigorous root systems would harvest the most water and would give the greatest yields. I do not mean to single Hurd out with this stricture. I have certainly made similar mistakes myself. But I want to emphasize that behaviour at one level of organisation cannot be easily extrapolated to a higher level. To revert to the language we used in discussing the example of the gas laws, it is extremely difficult to decide what are the most appropriate constraints to use when carrying out our integration (with 'constraints' corresponding to 'experimental conditions', and 'integration' to 'extrapolation').

Stomatal resistance

When water is limiting growth, the trade of CO_2 for water vapour by the plant becomes of great interest. Since the time of van den Honert's (1948) classic paper on resistances to the flow of water in the plant we have been strongly aware of the pivotal importance of the stomata in controlling this flow. Accordingly much work has been done on stomatal resistance, and it has become clear that although the transport of water vapour out of a leaf is controlled by only two resistances, that of the stomata and that of the boundary layer in the air surrounding the leaf, the transport of CO_2 into the leaf is controlled by three, the two involved for water, plus a third, associated with the movement of CO_2 through the cells of the leaf and the biochemical reactions within the chloroplasts. This third is known as the mesophyll resistance, and at first sight its presence would suggest that a plant would improve its balance of trade between CO_2 and water vapour if it increased its stomatal resistance; the relative importance of the hindering mesophyll resistance would thereby be decreased. This view resulted in a flurry of work on anti-transpirants and on attempts to control stomatal resistance by genetic means, the early promise of which has not been fulfilled. The early promise was, I think, largely due to the success of a self-fulfilling physiological prophecy similar to that in the previous example. The water-use-efficiency of a leaf in a temperature-controlled leaf chamber does indeed behave in the way we might expect. There are, however, two major difficulties associated with extrapolating these ideas to the field. The first is that the leaves of field-grown plants do not have their temperatures controlled, and when their stomata close the leaf-temperature rises, thus increasing the gradient in water vapour pressure between leaf and air. The result is that far from giving us the improvement in water-use-efficiency that we expected, we might even get a decrease in efficiency (Cowan and Troughton, 1971). The second difficulty is that even if the efficiency of the leaf were increased when the stomata closed, the continued respiration in the rest of the plant might well result in a decrease in efficiency based on the plant as a whole. There is also a third difficulty, rather different in kind from the other two, and resulting from the myopia generated by van den Honert's extremely influential paper. This difficulty is that in a field situation it appears that it is the control of leaf area, and not that of stomatal resistance, that may have the dominant influence on the exchange of CO_2 and water vapour (Legg *et al.*, 1979).

I have sought to illustrate, with these two examples, how difficult it is to extrapolate from work on one level of organisation to another level that is higher. The first example was concerned essentially with work at the whole-plant-level, the second, with work at the organ-level. How heroic it would be to extrapolate from work on tissues, cells, or enzymes, to the field-grown crop.

Such acts of heroism are rare, but do exist, and the rise and fall of proline as an indicator of drought tolerance is probably the best example.

As I remarked earlier, I believe that these extrapolations have been made particularly difficult by an initially too-fine dissection of the crop by the physioogist. There has to be a highly articulated cascade of understanding passing from the crop through each lower level of organization until we reach the one at which the physiologist happens to be working, if we are to have any chance at all of the ideas generated at that level being able to work their way, effectively, back to the level of the crop.

Coarse dissections of the crop

If, as I have just argued, the physiologist's dissection of the crop have been generally too fine to be of direct practical use to a breeder, what sorts of coarse dissection might be more suitable? Yield-component analysis is one, which has been popular since the time of Engledow and Wadham (1923). It involves viewing yield as the product of plants per unit area, ears per plant, grains per ear, and weight per grain. The enthusiasm with which this analysis was first applied was soon dampened when it became clear that there are generally strong negative correlations between the components (Adams, 1967): there is no point in trying to double the number of ears per plant if at the same time we halve the number of grains per ear. It is not until several days after anthesis that the number of grains per unit area is firmly set and is no longer susceptible to the plastic behaviour of the plant. Sensible yield-component analysis these days is restricted to studying only two components – the weight per grain, and the number of grains per unit area. The latter sets the potential yield of the crop, which is realised if conditions after anthesis are good enough for the grains to be completely filled. If conditions after anthesis are not good, the yield is largely determined by the amount of photosynthesis during grain filling, and the amount of stored assimilate that can be transferred to the grain. In these circumstances the two components are negatively correlated, and the analysis is not worthwhile. Nevertheless, it is probably worthwhile to aim for a large number of grains per unit area. The benefits of having a large potential yield would normally offset the risk of producing shrivelled grain.

This type of analysis, though conceptually very simple, provides an excellent framework within which to set physiological work on yield improvement that is being done at the level of the organ or of the whole plant. The focus of work on the rate of photosynthesis, for example, is greatly improved if we have some feeling for whether the photosynthate is influencing grain number or grain weight, and whether or not there is likely to be a negative correlation between the two. Similar remarks apply to the effects of drought on grain yield. The work of Fischer and Stockman (1980) on the effect of pre-anthesis shading on grain number, and that of Morgan (1980) on the effect of drought on grain number, provide good examples of nicely focussed physiology.

Another way of coarsely dissecting yield is to view it as the product of total dry matter (i.e., biological yield) and harvest index – the proportion of the total dry matter that is found in the grain. Donald (e.g. Donald and Hamblin, 1976) is the most eloquent champion of this view, and has argued convincingly that virtually all of the improvement in yield in wheat cultivars this century can be attributed to fortuitous improvements in harvest index that accompanied changes in other characters that breeders were seeking to introduce. He has argued further that, although harvest index depends on many environmental conditions, a plant that has a high harvest index when grown under good conditions in a glasshouse will tend to have a high harvest index when grown in plots in the field. The predictive value of grain yield in similar circumstances is virtually zero. Thus harvest index may provide a simple and effective criterion for selection during early generations. We will no doubt hear more of Professor Donald's views on harvest index later in the day.

My purpose in discussing harvest index is not primarily to focus on its possible role as an explicit criterion in breeding. Rather, I am interested in the framework it provides for designing physiological experiments that may reveal new characters that could be of use in a breeding programme. People who have attacked the idea of harvest index have argued that it is devoid of physiological meaning, that its correlation with yield is spurious, and that it is so susceptible to the vicissitudes of the environment that there is no point in spending time thinking about it. Where there are major environmental limitations, however, and particularly if there is drought, it is this very susceptibility of harvest index that makes it so interesting.

The biological yield of a water-limited wheat crop is largely explicable in terms of the available water supply and the efficiency with which that water is used in producing dry matter. There may be a chance in some environments, such as the Canadian one mentioned earlier, of improving the water supply; in many other environments, however, there appears to be little chance. There appear to be no consistent differences between cultivars in the efficiency with which they can use a given water supply in producing dry matter (Fischer and Turner, 1978), although, conceivably, partitioning of less assimilate into the root system could bring about some improvement (Passioura, 1977). Thus the production of biological yield during drought is reasonably well understood and seems to offer little scope for genetic improvement. Harvest index, on the other hand varies enormously in response to drought, ranging from zero to 0.5, and is much more susceptible to genetic improvement. The importance of earliness of maturity of wheat in the Australian environment, for example, can be conveniently discussed in terms of harvest index. Provided there have been no catastrophic effects on grain set, the harvest index of water-limited wheat depends strongly on the pattern of water use during the growing season. Where water-deficits are evenly distributed throughout the life of the crop they have little influence on harvest index, but where they occur predominantly during flowering and grain-filling they can greatly reduce it (Fischer and Turner, 1978).

The dissection of grain yield into the product of biological yield and harvest index thus provides an extremely useful bridge for relating grain yield during drought to fairly detailed physiological processes, insofar as these processes influence the pattern of water use by the crop. The growth of leaves during drought, the extraction by roots of water stored in the subsoil, the sensitivity of stomata to water deficits, all come into clearer focus, if we relate them first to their influence on the pattern of water use, before we try to relate them to grain yield.

The interaction between breeding and very detailed physiology

I have so far been arguing that, since breeding is concerned with improving performance at the community level, any physiological work that is likely to be a direct stimulus to the breeder must have fairly clear significance at the community level; it is pointless making a wild leap across a great chasm of understanding that may separate the behaviour of, say, a cell from that of a crop.

There is, however, a general exception to this remark. In a hierarchically organised system such as the wheat plant, there is an asymmetry of interaction between the various levels of organisation. The continued functioning of a low level does not necessarily require the proper functioning of a high level — a leaf, for example, remains alive after we cut it off no matter what we do to the rest of the plant. The proper functioning of a high level, however, requires the proper functioning of all lower levels. Anything that clearly upsets the performance of a plant at, say, the cellular level, is certain to upset the performance of the whole plant. A bacterial toxin that can affect any cell in the plant is a case in point, and one can envisage a successful selection procedure based on tissue culture that would improve resistance to such a toxin. We would be much less confident about using tissue culture to select for resistance to a vascular wilt disease such as that due to *Fusarium*, for that affects only certain types of highly differentiated cell. Similarly, selecting for salt tolerance at a cellular level is unlikely to be successful, for many different high-level processes (for example, the exclusion of salt at the endodermis of the root) are involved.

If cellular and sub-cellular physiology are going to have any major impact on wheat breeding, I believe it to be very unlikely that this would be through the discovery of important new characters that will influence grain yield. The greatest scope for their impact is in providing tools for speeding up that part of a breeder's programme that is effectively a random search.

Conclusion

I have tried to argue that we manage to make sense of complex, highly organised systems such as wheat plants by analysing them into conceptual layers that can only be properly understood in terms of concepts appropriate to each particular layer: just as the ordinary concept of pressure has no meaning at a molecular level except when a translation is provided by means of statistical mechanics, so the concept of maturity has no meaning at a cellular level, unless we have articulated the appropriate interactions between the cells. And just as the concept of pressure was discovered before its explanation in molecular terms, so was the concept of maturity discovered before its explanation in cellular terms. It seems that the properties of a given level of organisation have to be discovered by reflecting about that level, and that the Laplacean boast is quite empty, that a knowledge of the positions and velocities of all the molecules in the universe at any given time tells us the state of the universe for evermore. There is no point in knowing all about the behaviour of the molecules in a small segment of space if we do not thereby realise that those molecules constitute a wheat plant.

But though it seems that the properties of a given level can only be discovered by thinking about that level, it by no means follows that thinking about adjacent levels is a waste of time. On the contrary, it is the search for explanation, i.e. translation to a lower level, and the search for functional significance, i.e. translation to a higher level, that stimulates our imaginative thought. It does follow, however, that thought at a level remote from the one we are interested in is unlikely to provide much stimulus. Since breeding is concerned with performance in the field, the laboratory-bound physiologist is unlikely to make much useful impact on a breeder's thoughts unless the path to the field is very well marked. The dissection of yield into grain number and grain weight, or into biological yield and harvest index, is an important signpost on that path.

Perhaps the greatest chance that the physiologist has of successfully interacting with a breeder is to follow Donald's (1968) advice and to explore those characters in a plant that are likely to be beneficial in a crop of uniform genotype but deleterious in a natural community where different genotypes will be competing for limited resources. Dwarfism is a good example of this type of character, and one that has been extremely successful. Conservation in water-use is another, which is yet to prove its worth, but could be valuable where a crop is relying on a limited supply of stored water (Passioura, 1977). The attraction of conservative traits such as these is that they are likely to be rare in wild plants, and that they are not likely to last long in a breeder's plots unless he is specifically looking for them. They may therefore be rare even in modern cultivars. Traits that are likely to confer a competitive advantage on a plant in a competitive situation, such as a high photosynthetic rate per unit leaf area, are unlikely candidates for improving yield, for evolution over countless generations has probably perfected them.

Fischer (1977) has listed several physiological problems whose solution may be a stimulus to the breeder. The successes of the breeders in the 1950's and 1960's have certainly been a stimulus to the physiologist. But now that major breeding programmes seem to have lost their momentum, for the big improvements in yield in the 1960's have not been followed by similar improvements in the 1970's, it may be time to foster a close liaison between physiologist and breeder. Once the physiologist learns to work with more than one genotype at a time, and provided he maintains a well-worn path to the field, then the marriages between physiologists and breeders, whose proclamating banns are frequently seen nailed to the doors of grant-giving bodies, may prove to be abundantly fertile.

References

Adams, M.W. (1967) Basis of yield component compensation in crop plants with special reference to the field bean, *Phaseolus vulgaris. Crop Sci.* 7, 505—510.

Cowan, I.R. and Troughton, J.H. (1971) The relative role of stomata in transpiration and assimilation. *Planta (Berl.)* 97, 325—336.

Donald, C.M. (1968) The breeding of crop ideotypes. *Euphytica* 17, 385—403.

Donald, C.M. and Hamblin, J. (1966) The biological yield and harvest index of cereals as agronomic and plant breeding criteria. *Adv. Agron.* 28, 361—405.

Engledow, F.L. and Wadham, S.M. (1923) Investigations on yield in the cereals. *J. Agric. Sci.* 13, 390—439.

Evans, L.T. (1977) The plant physiologist as midwife. *Search (Syd.)* 8, 262—268.

Fischer, R.A. (1977) The physiology of yield improvement — past and future. 3rd International Congress of the Society for the Advancement of Breeding Researches in Asia and Oceania (SABRAO). Plant Breeding Papers, Vol. 1, 3(a)-1—13.

Fischer, R.A. and Stockman, Y.M. (1980) Kernel number per spike in wheat (*Triticum aestivum* L.): responses to preanthesis shading. *Aust. J. Plant Physiol.* **7**, 169–180.

Fischer, R.A. and Turner, N.C. (1978) Plant productivity in the arid and semi-arid zones. *Annu. Rev. Plant Physiol.* **29**, 277–317.

van den Honert, T.H. (1948) Water transport in plants as a catenary process. *Discuss. Faraday Soc.* **3**, 146–153.

Hurd, E.A. (1974) Phenotype and drought tolerance in wheat. *Agric. Meteorol.* **14**, 39–55.

Kramer, P.J. (1969) 'Plant and Soil Water Relationships : a Modern Synthesis.' McGraw-Hill, New York.

Legg, B.J., Day, W., Lawlor, D.W. and Parkinson, K.J. (1979) The effect of drought on barley growth : models and measurements showing the relative importance of leaf area and photosynthetic rate. *J. Agric. Sci.* **92**, 703–716.

Morgan, J.M. (1980) Possible role of abscisic acid in reducing seed set in water-stressed wheat plants. *Nature (Lond.)* **285**, 655–657.

Passioura, J.B. (1977) Grain yield, harvest index, and water use of wheat. *J. Aust. Inst. Agric. Sci.* **43**, 117–120.

Passioura, J.B. (1979) Accountability, philosophy and plant physiology. *Search (Syd.)* **10**, 347–350.

Rosen, R. (1969) Hierarchical organization in automata theoretic models of biological systems. In 'Hierarchical Structures'. (Eds. L.L. Whyte, A.G. Wilson, D. Wilson.) Amer. Elsevier, New York. pp. 179–199.

Sauerbeck, D. and Johnen, B. (1976) Der Umsatz von Pflanzenwurzeln im Laufe der Vegetationsperiode und dessen Beitrag zur 'Bodenatmung'. *Z. Pflanzenernaehr. Bodenkd.* **3**, 315–328.

13 YIELD IMPROVEMENT IN WHEAT: EMPIRICAL OR ANALYTICAL?

L.T. Evans

In a paper entitled 'The theory of plant breeding for yield', published in 1947, Otto Frankel wrote "The main increases in yield have been, and are being, achieved by overcoming limiting factors whose effects can be distinguished with a fair degree of certainty, rather than by assembling productivity genes, although the latter process is likely to have accompanied the former". He went on to point out that although some 'productivity genes' may have specific effects on observable yield components, many act indirectly through physiological processes which influence the overall level rather than the individual components of yield.

In these circumstances, the most effective way of increasing the yield potential of wheat might simply be to select empirically for yield itself. A more challenging and intellectually satisfying approach, which offers the possibility of more radical advance, would be to identify the physiological processes limiting grain yield and provide the plant breeder with practicable selection criteria for the improvement of the limiting processes. Such an analytical approach could confer greater specificity of aim and more discriminating selection, particularly in the early generations of a breeding program.

Yield criteria

The criterion of yield which is currently of most relevance is the amount of grain harvested per unit area of land per crop. However, different yield criteria have been of importance in the past history of wheat, while others may be more relevant in the future.

When our ancestors harvested grain from the wild cereal stands in the Fertile Crescent of the Middle East, the appropriate yield criterion was probably the amount of grain harvested per hour or per calorie of effort. Harlan (1967) gathered a kilogram of seed per hour from wild wheat stands in Turkey, which I estimate to have returned about 50 calories for each calorie spent on harvest. Under these conditions, plants with larger seeds, with more seeds per ear, and with more compact panicles or ears would be preferred. This last may be one reason why wheat and barley were domesticated earlier than oats, which Ladizinsky (1975) found to be more slowly harvested than wheat or barley from wild stands of all three cereals.

Once the sowing of wheat crops became established practice, a new criterion of yield became important, namely the ratio of seed harvested to seed sown. This is the measure of yield mentioned in the Bible and by Roman writers such as Columella, who refers to a fourfold return of wheat. In fact this was the measure of wheat yield applied throughout most of the history of the crop, and in Europe until relatively recently. One reason why Linnaeus student Peter Kalm was so impressed by maize when he first saw it in 1748 was that it often returned at least 300-fold, hence his description of it as 'the lazy man's grain', in contrast to wheat.

Given this emphasis on the seed return ratio, which is understandable when hungry families had to forego a substantial fraction of their harvest for the next sowing, wheat plants might well have been selected for abundant tillering and many grains per ear, as well as for strong seed retention. Seed dormancy mechanisms, prominent in some of the wild relatives of wheat, would have been eliminated in favour of uniform germination. Larger grains might also have been favoured to the extent that they gave larger seedlings better able to compete with weeds. Although the seed return ratios were often quite low, high ratios could be obtained with good husbandry: for example, wheat crops on good soils in Palestine in the period of the Mishnah returned up to 45-fold (Feliks, 1963).

As the pressure on arable land grew, grain yield per unit area per crop became a more common criterion of yield, and probably changed the selection pressures on wheat. Whereas many tillers and many seeds per plant maximized the seed return ratio, these characteristics are not necessarily conducive to high yield per hectare where, as Donald emphasizes in Chapter 14, plants which individually are weak competitors may give the most productive crops.

With further increase in the pressures on arable land, irrigation, fertilizers and other inputs are being used to an ever greater extent to substitute for land by raising the yield per hectare. As these other agricultural resources in turn become more limiting, the yield of wheat crops may be assessed not only in terms of land area but also in terms of water or phosphorus or energy used in their culture. Similarly, rising pressure on arable land is also leading to increased multiple cropping, particularly where this is made possible by irrigation and by warmer temperatures. Under these circumstances, grain yield per hectare per crop becomes less important than grain yield per hectare per day, and new selection pressures are introduced.

Thus, yield criteria for wheat crops in the future, as in the past, may differ from those currently emphasized, and as they change, so may the yield-limiting processes. But for the remainder of this paper we shall confine ourselves to yield per hectare per crop.

Changes in wheat yield per hectare

Yields of wheat in the Middle Ages (Figure 13.1) were often no higher than those of the wild cereal stands harvested by Zohary (1969), but they have risen progressively since then, and quite spectacularly over the last century.

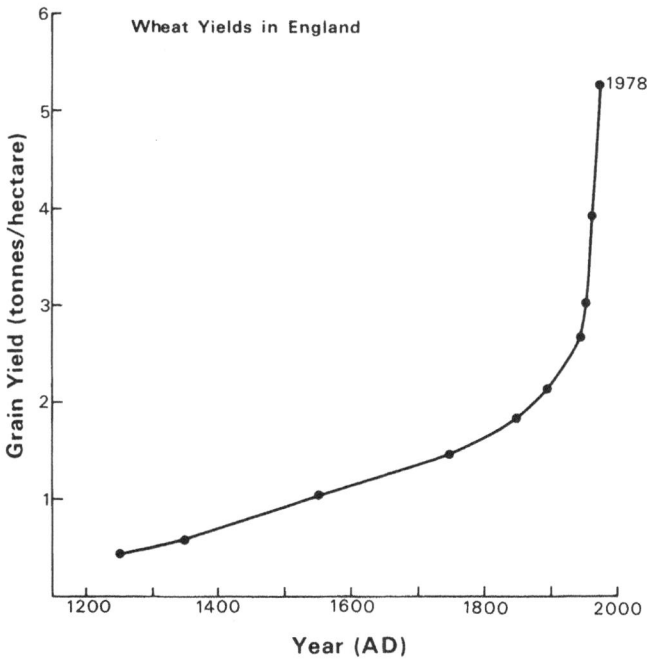

Fig. 13.1. Historical trend in the grain yield of wheat crops in England.

The eventual form of the yield curve is likely to be sigmoid, but so far there is no sign in the United Kingdom of an approaching asymptote, since wheat yields are still increasing linearly (Austin, 1978). Such increases in yield have been due partly to increased inputs and improved agronomic practices and partly to a variety of genetic changes effected by plant breeders, but these two sources of increase are not independent of one another. High input agriculture and continuous cropping may, for example, exacerbate disease and pest problems, thereby requiring more resistant lines to be bred. Likewise, the breeding and use of varieties which are weak competitors could not be contemplated prior to the development of effective herbicides. On the other hand, genetic changes such as the breeding of productive dwarf wheats made the heavier use of nitrogenous fertilizers economically worthwhile. So strong and varied are these interactions that there is little point in trying to partition advances in yield between genetic and agronomic components.

The genetic changes which influence wheat yields are of many kinds, and these need some discussion before we concentrate on what I shall call genetic yield potential.

1. First, there are the changes which conferred adaptedness to agriculture, such as non-shattering ears, non-dormant seeds and other characteristics selected for in the early stages of wheat domestication.

2. Secondly, there is the modification of those processes which control the timing of the life cycle of the crop as wheat spread from its hearth of domestication in the Middle East to higher latitudes in Europe and elsewhere, and then to lower latitudes in India, Australia and eventually even in the tropics. The regulatory processes which control inflorescence initiation and anthesis affect yield by determining not only whether the cultivar can flower and when it does so — which may be of over-riding importance in many environments — but also how large the ears will be. The yield potential of cultivars can be assessed, therefore, only in environments to which they are well adapted.

3. Yield may also depend to a considerable degree on the ability of cultivars to survive or escape the adverse effects of drought, heat and cold. Many other characteristics besides the timing of the life cycle help to reduce the adverse effects of these environmental stresses but some of these may have adverse effects on yield potential under favourable conditions.

4. Selection for baking and bread making qualities may also have indirect effects on the yield of wheat. For example, the higher energy cost for the synthesis of protein compared with that of starch on a weight for weight basis means that high quality wheats will suffer a yield penalty if yield is limited by the supply of energy rather than of carbon. Also, the higher protein content of the grain may, in some cases, require earlier or greater remobilization of protein from leaves, thereby reducing their capacity to generate further assimilate for grain growth. Both these and other factors could contribute to the yield penalty usually associated with high quality wheats and evident in comparisons of yield potential among English winter wheat varieties (Figure 13.2).

5. Wheat yield can be profoundly influenced by selection for resistance to various pests and diseases. Under low input conditions, in which much of the world's wheat is grown, the incorporation of genetic resistance to major pests and diseases is crucial because control by the use of insecticides and fungicides may not be economic. Under high input conditions, where the incidence of pests and diseases may be enhanced, genetic resistance is preferable to the use of pesticides, and may often be needed to supplement it. However, given the rate at which new biotypes of pests and diseases arise to bypass both varietal resistance and pesticides when these are employed over large areas, the concentration

of plant breeding effort on selection for genetic resistance can divert attention from 'the assembling of productivity genes'.

6. 'Resistance' to the increasing array of fertilizers and agrichemicals used on wheat crops is also needed. The best known example of this is the selection of lodging-resistant shorter-stemmed wheats for use with heavy applications of nitrogenous fertilizers, but insensitivity to widely used herbicides and pesticides is also required (Gressel, 1978). Seedling insensitivity to residual herbicides may also be important for wheat grown under minimum tillage.

These six kinds of genetic modification by plant breeding all influence wheat yield, some of them profoundly. They are dealt with more comprehensively in other chapters of this book, but I have listed them here in order to distinguish them from a further kind of genetic modification, namely increased yield potential. This may be defined as the yield of a cultivar grown in environments to which it is adapted, with nutrients and water non-limiting and with pests, diseases, weeds, lodging and other stresses effectively controlled. Only under such conditions are we able to assess progress in the assembling of 'productivity genes' as distinct from genes for adaptation to environment, adaptedness to modern agronomy and resistance to pests and diseases. Moreover, wheat crops are now grown under high input agronomy in many countries.

Rate of increase in genetic yield potential

Increase in the genetic yield potential of wheat can be assessed, to some extent, from records of national yield trials conducted over a period of years, provided the successive standard varieties overlap long enough for their relative performance to be adequately compared. MacKey (1979) presents the results of Swedish trials with both spring and winter wheat, each series beginning with an 'unbred, indigenous variety'. His data for winter wheat can be fitted quite well by an exponential curve increasing at a rate of 0.42% per year, whereas the Swedish national average yield has increased at 3.7% per year over the same period.

A comparable analysis of the results of winter wheat trials conducted by the National Institute of Agricultural Botany (NIAB) in the United Kingdom is presented in Figure 13.2. Since the introduction of the variety 'Little Joss' in 1908, the yield potential of winter wheats has been increased by more than 50%, with the high quality baking wheat varieties lagging behind. The increase in yield potential has been somewhat irregular, with a sharp rise occurring in the 1970's with the introduction of semi-dwarf varieties. The overall rate of increase for the higher yielding varieties has been 0.87% per year, while that for the varieties of high baking quality has been 0.68% per year.

When the results from NIAB trials with spring wheat varieties are compiled in this way, they fit closely to an exponential curve with a rate of increase of

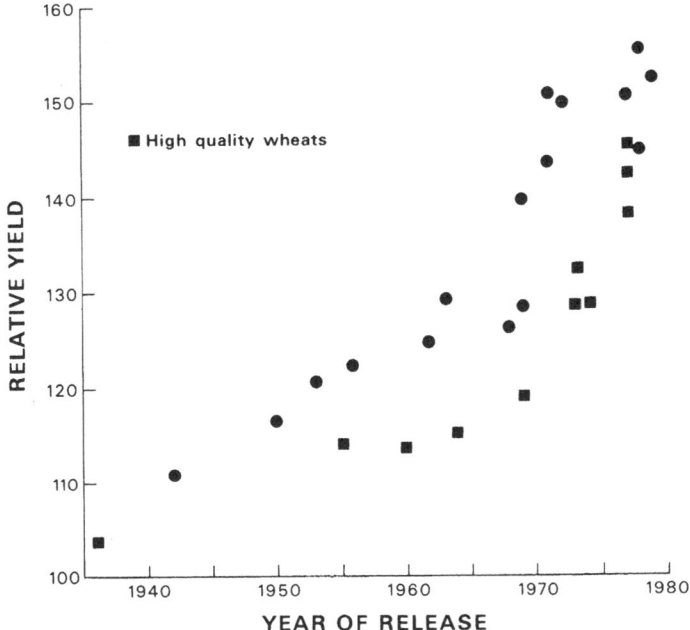

Fig. 13.2. Estimated increase in the yield potential (relative to Little Joss, released in 1908, = 100) of varieties of winter wheat released in the United Kingdom over the last 50 years, based on the results of annual trials conducted by the National Institute of Agricultural Botany.

2.23% per year since 1950 (Evans, 1980). The much higher rate of increase for the spring wheat varieties may be due to the fact that less attention was given to their improvement in earlier years than to that of the winter wheats. However, for neither winter nor spring wheat varieties is there any indication of an approaching plateau in yield potential.

Such trial results have the advantage that varieties are compared under the agronomic conditions for which they were bred. Their disadvantage is that they may over-estimate the rate of increase in yield potential if the outgoing standard variety begins to fail in its resistance to disease before the new standard takes over. However, only occasional examples of this were evident in the comparative NIAB yield ratings. A more direct comparison of the relative yield potential of winter wheat varieties bred in Cambridge over the last 70 years has been made by Austin *et al.* (1980) at both low and high levels of fertilizer application, the varieties being grown under conditions where diseases were controlled by spraying and where lodging was prevented by the use of mesh. Even the oldest variety included in the experiment, Little Joss, gave a grain yield of 5.9 t ha^{-1} (at 12% moisture) at the higher level of fertility, but the highest yielding recent variety surpassed that by 45%. The overall increase in yield potential indicated by this experiment, at both levels of fertility, was therefore somewhat less than that suggested by the NIAB trial results in Figure 13.2, but the cool, cloudy growing season in 1978 may have led to some under-estimation of the yield potential of the newer varieties.

The contribution made by these increases in yield potential to the rise in wheat yields in England and Wales between 1947 and 1975 has been estimated by Silvey (1978), taking account of the extent to which the various cultivars have been grown. Overall, about half of the 2-3% per year increase in national yield has been due to greater genetic yield potential, but this proportion has risen strikingly throughout the period covered by Silvey's analysis.

Comparable data for wheat crops in other countries are less comprehensive. Wheat yields in New York State have increased steadily since 1935 and, from the results of his 'living museum' of New York varieties, Jensen (1978) concludes that almost exactly half of this increase is due to the 47% greater yield potential of modern varieties such as cv. Yorkstar compared with cv. Honor released in 1920. In New Zealand, where Otto Frankel played a major part in the breeding of many improved varieties of wheat, Smith (1971) also estimates that about half of the national yield increase has been due to genetic improvement and half to better crop husbandry. In Mexico, on the other hand, comparative yields of varieties introduced between 1950 and 1973, and grown under high input conditions by Fischer and Wall (1976), indicate an increase in yield potential of about 0.9% per year, whereas Mexican wheat yields have risen by 5.2% per year. In Australia, wheat yields have risen more slowly and irregularly, but the rate of increase in yield potential has been 0.6% per year according to Donald (1965). An experiment by Davidson and Birch (1980) with 61 varieties used in Australia over the last century suggests that the rise in yield potential has been rather irregular, most of it occurring in two steps, with the introduction of Gabo in 1945 and of Egret in 1973.

One question with important implications for the formulation of national policies for food reserves is whether crop yields become more or less variable as they rise to higher levels. For the wheat crop in the USA, Luttrell and Gilbert's (1976) analysis indicated that the incidence of yields 5 or 10% below trend has diminished somewhat in recent years. Stanhill's (1976) long term analysis of wheat yields in England indicated that there has been no increase in the coefficient of variability as yields have risen, and this appears to be true also for wheat yields in Argentina, Australia, Canada, India and the USA (Waggoner, 1979).

The physiological basis of increased yield potential

The results presented in the previous section indicate that a substantial increase in the yield potential of wheat has been achieved by plant breeders in several countries, through empirical selection. The nature of this increase will now be examined, and some of the ways in which it might be further enhanced. The effects of changes in stature, in leaf inclination and size, and in tillering are discussed more fully by Donald in Chapter 14.

(a) Evolutionary perspective

Physiological comparisons between lower and higher yielding varieties of wheat have long been made, but only in the last ten years or so have these been broadened to include the wild relatives and presumed progenitors, although comparable cytogenetic studies have provided much of our understanding of crop evolution.

In the course of domestication and improvement of wheat there has been a substantial increase in grain size. Leaf size has increased in parallel, from the wild diploids to the cultivated hexaploids (Evans and Dunstone, 1970), but this may be largely associated with the increasing ploidy since comparable increases have not occurred during the domestication of several other crop plants, such as rice. Despite the increase in leaf area, there has been no increase in the relative growth rate of seedlings (Evans and Dunstone, 1970; Khan and Tsunoda, 1970b).

Nor has there been any increase in leaf photosynthetic rates. In fact, the highest rates - very high indeed for a C_3 plant - occur in the wild diploid *T. boeoticum*, and maximum flag leaf rates have tended to fall in the course of wheat evolution (Evans and Dunstone, 1970; Khan and Tsunoda, 1970a). However, there has been no shift in the relation between photosynthesis and photorespiration, nor in that between the stomatal and the residual components of leaf resistance to CO_2 assimilation (Dunstone *et al.*, 1973). The lower maximum rate of flag leaf photosynthesis in the modern wheats is associated with greater size of the leaf and of the mesophyll cells (Dunstone and Evans, 1974). Miginiac-Maslow *et al.* (1979) also found a negative relation between protoplast diameter and Hill reaction activity among the wheats. However high rates of Hill reaction activity were also found in isolated chloroplasts of the diploid wheats (Zelenskii *et al.*, 1978). Thus, the faster photosynthesis in the more primitive wheats cannot be ascribed wholly to greater surface/volume ratios of the mesophyll cells.

In the course of wheat evolution there has not been much increase in the number of spikelets per ear, but the number of grains per spikelet has increased, as well as grain size (Evans and Dunstone, 1970). Larger grains are associated with both faster and longer grain growth (Sofield *et al.*, 1977a). The faster growth of more grains requires a faster rate of translocation of photosynthetic assimilates to each ear, and the cross-sectional area of phloem in the culm has increased in parallel with the increased need for imported assimilates in modern hexaploid cultivars compared with wild diploids (Evans *et al.*, 1970). A greater proportion of ^{14}C-labelled photosynthate is translocated from the leaves to the grains in modern wheats (Evans and Dunstone, 1970).

Thus, the evolution and improvement of wheat has not been accompanied by any increase in photosynthetic capacity or growth rate but rather by selection for greater size and number of kernels per ear and of the vascular bundles supplying them. The accumulated dry weight of the crops has not been increased, but rather the proportion of it which is allocated to the harvested grains.

(b) Denser crops

With higher inputs of fertilizer and better control of diseases and weeds, denser stands with more ears and grains m^{-2} can be supported through to maturity, with a consequent increase in the potential storage capacity and yield of the crop. Fischer *et al.* (1977) and Evans (1978) have shown the close relation between yield and ear number up to 1,000 m^{-2} or grain number up to 30 000 m^{-2}, as determined largely by irradiance. At such densities, the leaf area index (LAI) of the crop can reach very high values before anthesis.

Wheat canopies with an LAI of 4-6 are sufficient for almost complete interception of photosynthetically-active radiation, and for some time it was considered that higher LAI values would reduce net photosynthesis by the crop due to respiration by the lower shaded leaves exceeding their photosynthesis. However, it is now known that both dark respiration and net photosynthesis reach a plateau as LAI increases (King and Evans, 1967), as does crop growth rate (Watson *et al.*, 1963; Stoy, 1965), due to the respiration rate of the lower shaded leaves being reduced in parallel with their rate of photosynthesis. However, although there seems to be no need to reduce peak LAI values - except possibly under conditions where they are associated with greater incidence of diseases - there may well be advantages in modifying leaf dimensions and inclination for better performance in denser stands, as discussed in the next chapter.

(c) Faster growth and development

Greater use of fertilizers and better control of diseases and weeds allows crops to grow more rapidly.. Varietal comparisons in both the United Kingdom (Austin *et al.*, 1980) and Australia (Davidson and Birch, 1980) suggest that modern varieties are slightly faster developing, but there are often major environmental constraints on the extent to which crop development can be accelerated.

Faster establishment and vegetative growth allow the young crop to intercept more radiation, but this advantage is lost once the canopy has closed. Faster development through the vegetative stage may even be disadvantageous, by reducing grain weight per ear to some extent (Warrington *et al.*, 1977). A vernalization requirement, on the other hand, may increase the number of spikelets and grains per ear by delaying inflorescence initiation (Rawson, 1970).

Rapid development in the reproductive phase, between ear initiation and anthesis, has an even more drastic effect on yield components in wheat. Higher temperatures during this stage greatly reduce grain number and weight per ear (Warrington *et al.*, 1977). Rawson (1970) found that the acceleration of terminal spikelet formation by exposure to longer days also reduced spikelet and grain number. Likewise, varieties with a shorter interval between the initiation of double ridges and of the terminal spikelet had fewer spikelets and usually fewer grains per ear.

Considerable genetic variation also exists in the duration of the final grain growth phase. The greater the duration, whether genetic or achieved by lower temperatures, the greater is the weight per grain (Spiertz, 1974; Sofield *et al.*, 1977a). In conditions where grain growth is not brought to a halt by environmental stress, it may well be feasible to select for a longer duration of grain growth since this may be terminated not so much by lack of assimilates as by the deposition of lipids in the chalazal zone of entry into the grain (Sofield *et al.*, 1977b).

Thus, in terms of yield per crop, selection for faster development at any phase of the life cycle of wheat is likely to have an adverse effect on yield potential, although it may be valuable in improving the adaptation of a cultivar to the seasonal cycle.

(d) Enhanced longevity

Under low input agricultural conditions, most of the nitrogen (N) and phosphorus (P) in the mature wheat crop is taken up before anthesis, and most (80-90%) of that in the leaves is then remobilized into the grain by maturity (Williams, 1955). Such remobilization is associated with a fall in the content of ribulose-1,5-bisphosphate carboxylase and other photosynthetic enzymes (Patterson *et al.*, 1980), as well as of structural proteins, a loss of photosynthetic capacity, and senescence of the leaves, usually but not always from below upwards. In these circumstances, grain growth depends not only on remobilized N and P but also to a considerable extent on remobilized reserves of carbohydrate. Where N and P supplies from the soil are limiting, it could be counterproductive to try to enhance leaf longevity.

With higher levels and later applications of N and P fertilizers, however, nutrient uptake can continue after anthesis, with the consequence that remobilization of nutrients from the upper leaves need not be so rapid or extensive. Nevertheless, it may still occur (e.g. Williams, 1955), even when the ears are removed (Patterson and Brun, 1980). However, the timing and pattern of senescence varies considerably between varieties, and it should be possible to select for enhanced leaf longevity. Thomas *et al.* (1978) delayed flag leaf senescence by the application of nitrogenous fertilizers after ear emergence, but grain growth did not increase proportionately. However, selection for enhanced leaf longevity might increase yield potential if it was coupled with selection for greater duration of grain growth.

(e) Enhanced harvest index

The main conclusion from our evolutionary perspective on wheat was that the increase in yield potential has come mainly from a greater allocation of crop dry matter to the harvested grains. As van Dobben (1962) first pointed out, there has been little change in total crop dry weight but a progressive increase in total grain weight m^{-2}, and therefore in harvest index. Among British wheat varieties introduced over the last 70 years, Austin *et al.* (1980) found the harvest index to have risen from 34 to 50%, and they estimate that it could probably continue to rise to about 62%. A comparable rise has occurred in other wheat varieties (Evans, 1980).

Further increase in harvest index will probably be achieved indirectly, as it has in the past, by empirical selection for yield. Fischer and Kertesz (1976) found the harvest index of spaced plants to be more closely correlated with plot yields than were the yields of spaced plants, and suggested that shoot harvest indices would be a useful selection criterion in the early spaced-plant generations of a breeding program. However, the harvest index of tillers varies somewhat with their order of origin and with growing conditions (Davidson and Birch, 1978), which could complicate its use as a selection criterion.

(f) Enhanced competitiveness by the ear

Further increase in the harvest index of wheat is likely, on past trends, to take place at the expense of investment in organs other than the grain, and some of the processes involved will now be considered. By the time grain growth begins in wheat, root and shoot growth have largely ceased. Stem weight may continue to increase in some varieties (e.g. Wardlaw, 1970; Rawson and Evans, 1971), mostly due to the temporary accumulation of carbohydrate reserves. Some late tillering may also occur, but for most of the post-anthesis period the grains are the dominant sink for current assimilates and reserves.

Prior to anthesis, however, the developing ear must compete for assimilate with several strong sinks. The roots are still growing, although at a slower rate (Connor, 1975). Tiller numbers may rise substantially for several weeks after floral initiation in many varieties (Evans *et al.*, 1975). Within the shoot, the developing ear must also compete with growth of the uppermost leaves and with the major phase of stem growth (Fischer and Stockman, 1980), during the very period when the yield potential of the ear is being defined by the differentiation first of spikelets and then of florets. Evidence of competition between flag leaf growth and the developing ear at the stage when the number of spikelets is being determined was found by Rawson (1970). At later stages of its development the ear is particularly subject to competition with the rapidly growing stem, and shading treatments reduce the allocation of assimilates to it disproportionately while it is a small sink but not later on when its growth rate almost equals that of the stem (Fischer and Stockman, 1980).

This analysis would suggest that the determination of yield potential may be particularly sensitive to the supply of assimilate during the reproductive stage, especially during the early development of the ear. In fact, this is the stage at which the correlation between daily irradiance and grain yield is highest (Willey and Holliday, 1971; Fischer, 1975; Evans, 1978), and when yield is most responsive to greater CO_2 concentration (Fischer and Aguilar, 1976).

One key to increased yield potential lies, therefore, in the greater ability of the differentiating ear to compete with other organs for assimilates. Such competition is influenced by the relative size of the competing organs, by their relative distance from the source of photosynthate, and by the relative directness and capacity of their vascular connections to it. Cook and Evans (1978) tried to assess the importance of these factors in wheat by varying the number of grains in two ears competing for assimilates from one leaf at various positions. The results gave clear evidence of the advantage of proximity to the source leaf, and more ambiguous evidence of the advantage of greater sink size.

In recent experiments, I have used a more satisfactory experimental system in which both the only source of photosynthate (a single awn on the basal floret of a central spikelet) and the competing sinks (grains in two spikelets) are within one ear and only a few cm apart. One advantage of this system is that both attached and detached ears can be used. Some results with the variety Siete Cerros are illustrated in Figure 13.3, and comparable results were obtained with cv. Highbury. About 80% of the ^{14}C fixed by the awn was exported to the competing grains, which is in itself interesting since no such export to other spikelets occurs in intact ears, indicating that distribution patterns are largely determined by where the nearest dominant sinks are. In these experiments, a strong bias in favour of upward movement of assimilates is evident (Treatment

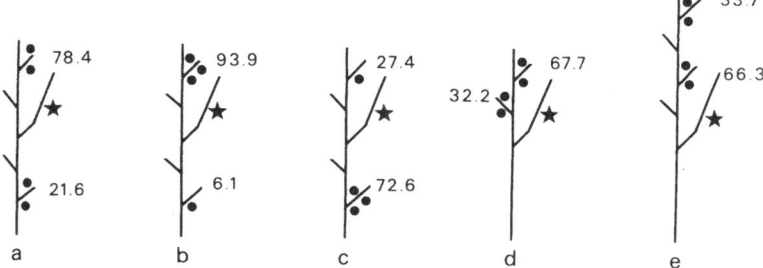

Fig. 13.3. Percentage distribution of ^{14}C, fixed by a centrally placed awn, between grains in two competing spikelets of ears of Siete Cerros wheat. The awn (starred) was the only source of current photosynthate, and only 4 evenly sized grains were left in each ear.

a). When there were more grains above the fed awn (Treatment b), this upward bias was accentuated even further, but it was completely overcome when there were more grains below it (Treatment c). When all grains were above the fed awn, those which were closer gained twice as much activity as those twice as far away on the same side of the ear (Treatment e), and in other treatments with equally sized competing sinks, the share received by the more distant sink fell progressively as the distance between the sinks increased. However, grains on the opposite side of the ear, presumably with less direct vascular connections to the fed awn, were at a disadvantage even when closer (Treatment d). These few treatments indicate, therefore, that where sinks are competing for assimilate, it is advantageous to be larger, closer to the source, and with more direct connections to it. In more recent experiments with Mary Cook under somewhat different conditions, these results have been extended and confirmed.

In these experiments, the competing sinks are similar in kind, whereas the differentiating inflorescence and the elongating stem are very different competitors. Assimilates moving towards the ear must pass through the upper internodes of the stem. This stem, by virtue of its greater proximity and size and better developed vascular system will presumably have an advantage in the competition for assimilates, at least until the ear is growing almost as rapidly as the stem. This interpretation is supported by the results of Fischer and Stockman (1980), and would imply that the reduction of assimilate requirements for stem growth during ear development should, provided other sinks are not increased, permit the enhancement of yield potential regardless of the genetic or other cause of dwarfing. In fact, the results of Austin *et al.* (1980) suggest that the enhancement of yield potential has been equivalent to the reduction in shoot weight.

(g) Enhanced photosynthesis

Just because there has been no increase in maximum photosynthetic rate per unit leaf area so far in wheat improvement is not a sufficient reason for discounting the possibility in future. The absence of any increase to date may imply that photosynthesis has already been perfected in evolution, as Passioura suggests in Chapter 12, or it may have resulted from counter-productive effects associated with selection for faster photosynthesis. When photosynthesis is increased by other means, such as CO_2 enrichment, there can be an increase in the yield of wheat crops grown under controlled conditions at higher temperatures (Krenzer and Moss, 1975; Gifford, 1977) and in the field under some conditions (e.g. Krenzer and Moss, 1975; Fischer and Aguilar, 1976) but not under others. Although the effect of CO_2 enhancement on yield is variable, being most pronounced under low light but not low temperature conditions, the enhancement is strong enough for us to have expected at least some indirect selection for higher photosynthetic rate to have occurred during wheat improvement.

One reason why it has not could be that the photosynthetic rate was more readily enhanced by the use of nitrogenous fertilizers. This effect is striking in rice, but in wheat it is not always apparent. Osman *et al.* (1977) obtained a 3-fold enhancement of photosynthetic rate per unit leaf area with increase in the level of nitrogenous fertilizer, but the Rothamsted group consistently find increase in leaf area rather than in photosynthetic rate at higher levels of N (Thomas *et al.*, 1978; Thorne *et al.*, 1979).

Another reason why photosynthetic rate has not been increased may be a negative relation between it and leaf area. This was apparent across wheat species, from wild diploid to modern hexaploid (Evans and Dunstone, 1970), and has since been found in varietal comparisons (Gale *et al.*, 1974; Planchon, 1979). It is presumably associated with the need to partition the materials for leaf growth into either greater area or greater investment per unit leaf area, which would make possible a higher photosynthetic rate. Large leaves are not needed in the later stages of crop development, when the canopy has closed, but they may be advantageous in the earlier stages by hastening canopy closure. A successful cultivar may, therefore, involve a compromise between the need for larger leaves in early growth and for higher photosynthetic rate in its upper leaves.

Another compromise to be made may be that between the maximum rate and the duration of photosynthetic activity. The wild diploid wheats have the highest rates but the most rapid senescence of their flag leaves, whereas modern cultivars retain their photosynthetic activity for much longer although their early peak rates are lower. Thus, whereas comparisons near anthesis reveal the highest rates in the wild diploids, several weeks after anthesis the photosynthetic rankings are reversed. The relation between photosynthetic rate and yield therefore depends very much on when photosynthesis is measured.

Conclusion

In the later stages of a plant breeding program, when selection for yield potential can be based on plot yields, such empirical selection is likely to be most effective for this characteristic. In the earlier generations, however, when spaced plants must be heavily culled, physiological criteria and techniques might permit more discriminating selection by plant breeders.

In the next chapfer, Donald makes the point that for a wheat crop to be high yielding, the individual plants making it up should be weak competitors. If his conclusion is correct, it follows that the selection of vigorous and productive plants in the early generation rows may actually be working against selection in later generations on the basis of plot yield, and it is important to establish the generality of Donald's argument.

The physiological perspective presented above supports Donald's view. The rise in grain yield potential has been at the expense of investment in other organs, and associated with reduced stem height, reduced leafiness and reduced production of unfertile tillers. Plants improved in these ways are bound to be weaker competitors in some situations. Their selection is made possible, however, by increased agronomic support for the crop. For example, where weeds are effectively controlled by herbicides, stems need not be so tall nor leaves so large and light-excluding; where water and nutrients are provided, root systems need not be so extensive; where pests, diseases and stresses are controlled, reserves to aid recovery need not be so large, and can be mobilized into the grain. Further raising of the level of agronomic support for the crop should make possible the selection of plants with an even higher harvest index, and these will inevitably be weaker competitors. The meek shall inherit the croplands of the earth.

However, this does not mean that varieties with increased rates of photosynthesis, growth or other processes should not be sought. Physiological analysis should be of value here in defining which yield limiting processes should be augmented. There has, for example, been much emphasis on the need to raise the rate of photosynthesis, and portable and rapid methods of selecting for increased leaf permeability and photosynthesis have been developed (e.g. Shimshi and Ephrat, 1975). But for such selection to be effective in raising yield, it may be necessary to break certain negative associations, such as that between maximum photosynthetic rate and leaf size or duration in some materials. Another example of negative associations is to be seen in the attempt by Jones (1977) to reduce crop water use by selecting for lower stomatal frequency. It turned out, however, that this characteristic was associated with greater stomatal size, cell size, leaf area and leaf longevity, and resulted in greater rather than less use of water.

This example highlights a need that is likely to become increasingly important as advances in recombinant DNA and other techniques make it more feasible to identify and recombine individual genes in a plant breeding program, namely the need to be able to predict the net physiological effect of transferring what Frankel (1947) referred to as 'productivity genes'.

The essence of wheat evolution to date is that the directly productive processes, such as photosynthesis, have been relatively conservative. It is the regulatory processes, such as those governing the timing of the life cycle or the control of plant form through the allocation process, that have been modified empirically to increase yield potential. We know a little of the subtlety and complexity of these regulatory processes in the wheat plant, but not nearly enough to be sure that the analytical approach to yield improvement combined with the utilization of new techniques for gene transfer will prove to be more effective than empirical selection in raising the yield of wheat.

References

Austin, R.B. (1978) Actual and potential yields of wheat and barley in the United Kingdom. *ADAS (Agric. Dev. Advis. Serv.) Q. Rev.* **29**, 76–87.

Austin, R.B., Bingham, J., Blackwell, R.D., Evans, L.T., Ford, M.A. Morgan, C.L., and Taylor M. (1980) Genetic improvements in winter wheat yields since 1900 and associated physiological changes. *J. Agric. Sci.* **94**, 675–689.

Connor, D.J. (1975) Growth, water relations and yield of wheat. *Aust. J. Plant Physiol.* **2**, 353–366.

Cook M.G., and Evans, L.T. (1978) Effect of relative size and distance of competing sinks on the distribution of photosynthetic assimilates in wheat. *Aust. J. Plant Physiol.* **5**, 495–509.

Davidson, J.L., and J.W. Birch (1978) Responses of a standard Australian and a Mexican wheat to temperature and water stress. *Aust. J. Agric. Res.* **29**, 1091–1106.

Davidson, J.L., and Birch, J.W. (1980) Yield trends from Australian wheats. Proc. Aust. Agronomy Conf. Qld. Agric. Coll. p. 231.

Dobben, W.H. van (1962) Influence of temperature and light conditions on dry matter distribution, development rate, and yield of arable crops. *Neth. J. Agric. Sci.* **10**, 377–389.

Donald, C.M. (1965) The progress of Australian agriculture and the role of pastures in environmental change. *Aust. J. Sci.* **27**, 187–198.

Dunstone, R.L., and Evans, L.T. (1974) Role of changes in cell size in the evolution of wheat. *Aust. J. Plant Physiol.* 1, 157–165.

Dunstone, R.L., Gifford, R.M. and Evans, L.T. (1973) Photosynthetic characteristics of modern and primitive wheat species in relation to ontogeny and adaptation to light. *Aust. J. Biol. Sci.* 26, 295–307.

Evans, L.T. (1978) The influence of irradiance before and after anthesis on grain yield and its components in microcrops of wheat grown in a constant daylength and temperature regime. *Field Crops Res.* 1, 5–19.

Evans, L.T. (1980) Response to challenge: William Farrer and the making of wheats. *J. Aust. Inst. Agric. Sci.* 46, 3–13.

Evans, L.T., and Dustone, R.L. (1970) Some physiological aspects of evolution in wheat. *Aust. J. Biol. Sci.* 23, 725–741.

Evans, L.T., Dunstone, R.L. Rawson, H.M. and Williams, R.F. (1970) The phloem of the wheat stem in relation to requirements for assimilate by the ear. *Aust. J. Biol. Sci.* 23, 743–752.

Evans, L.T., Wardlaw, I.F. and Fischer, R.A. (1975) Wheat. In 'Crop Physiology: Some Case Histories'. (Ed. L.T. Evans.) Cambridge Univ. Press, Cambridge. pp. 101–149.

Feliks, J. (1963) 'Agriculture in Palestine in the Period of the Mishna and Talmud. ' Magnes Press, Jerusalem.

Fischer, R.A. (1975) Yield potential in a dwarf spring wheat and the effect of shading. *Crop Sci.* 15, 607–613.

Fischer, R.A., and Aguilar, I. (1976) Yield potential in a dwarf spring wheat and the effect of carbon dioxide fertilization. *Agron. J.* 68, 749–752.

Fischer, R.A., Aguilar, I., and Laing, D.R. (1977) Post anthesis sink size in a high-yielding dwarf wheat: yield responses to grain number. *Aust. J. Agric. Res.* 28, 165–175.

Fischer, R.A., and Kertesz, Z. (1976) Harvest index in spaced populations and grain weight in microplots as indicators of yielding ability in spring wheat. *Crop Sci.* 16, 55–59.

Fischer, R.A., and Stockman. Y.M. (1980) Kernel number per spike in wheat (*Triticum aestivum* L.): responses to pre-anthesis shading. *Aust. J. Plant Physiol.* 7, 169–180.

Fischer, R.A., and Wall, P.C. (1976) Wheat breeding in Mexico and yield increases. *J. Aust. Inst. Agric. Sci.* 42, 139–148.

Frankel, O.H. (1947) The theory of plant breeding for yield. *Heredity* 1, 109–120.

Gale, M.D., Edrich, J., and Lupton, F.G.H. (1974) Photosynthetic rates and the effects of applied gibberellin in some dwarf, semi-dwarf and tall wheat varieties (*Triticum aestivum*). *J. Agric. Sci.* 83, 43–46.

Gifford, R.M. (1977) Growth pattern, carbon dioxide exchange and dry weight distribution in wheat growing under different photosynthetic environments. *Aust. J. Plant Physiol.* 4, 99–110.

Gressel, J. (1978) Genetic herbicide resistance : projections on appearance in weeds and breeding for it in crops. In 'Plant Regulation and World Agriculture.' (Ed. T.K. Scott.) Plenum, N.Y. pp. 85–109.

Harlan, J.R. (1967) A wild wheat harvest in Turkey. *Archaeol.* 20, 197–201.

Jensen, N.F. (1978) Limits to growth in world food production. *Science* 201, 317–320.

Jones, H.G. (1977) Transpiration in barley lines with differing stomatal frequencies. *J. Exp. Bot.* 28, 162–168.

Khan, M.A., and Tsunoda, S. (1970a) Evolutionary trends in leaf photosynthesis and related leaf characters among cultivated wheat species and its wild relatives. *Jpn. J. Breed.* 20, 133–140.

Khan, M.A., and Tsunoda, S. (1970b) Growth analysis of cultivated wheat species and their wild relatives with special reference to dry matter distribution among different plant organs and to leaf area expansion. *Tohoku J. Agric. Res.* 21, 47–59.

King, R.W., and Evans, L.T. (1967) Photosynthesis in artificial communities of wheat, lucerne, and subterranean clover plants. *Aust. J. Biol. Sci.* 20, 623–635.

Krenzer, E.G., and Moss, D.N. (1975) Carbon dioxide enrichment effects upon yield and yield components in wheat. *Crop Sci.* 15, 71–74.

Ladizinsky, G. (1975) Collection of wild cereals in the Upper Jordan Valley. *Econ. Bot.* 29, 264–267.

Luttrell, C.B., and Gilbert, R.A. (1976) Crop yields : random, cyclical or bunchy? *Am. J. Agric. Econ.* 58, 521–531.

MacKey, J. (1979) Genetic potentials for improved yield. Proc. Workshop on Agricultural Potentiality Directed by Nutritional Needs, Akad. Kiado, Budapest. (Ed. S. Rajki.) pp. 121–143.

Migniniac-Maslow, M., Hoarau, A. and Moyse, A. (1979) Hill reaction studies with protoplasts from cultivated wheats and their wild relatives. *Z. Pflanzenphysiol.* 95, 95–104.

Osman, A.M., Goodman, P.J. and Cooper, J.P. (1977) The effects of nitrogen, phosphorus and potassium on rates of growth and photosynthesis of wheat. *Photosynthetica* 11, 66–75.

Patterson, T.G., and Brun, W.A. (1980) Influence of sink removal on the senescence pattern of wheat. *Crop Sci.* 20, 19–23.

Patterson, T.G., Moss, D.N. and Brun, W.A. (1980) Enzymatic changes during the senescence of field-grown wheat. *Crop Sci.* **20**, 15–18.

Planchon, C. (1979) Photosynthesis, transpiration, resistance to CO_2 transfer, and water efficiency of flag leaf of bread wheat, durum wheat and Triticale. *Euphytica* **28**, 403–408.

Rawson, H.M. (1970) Spikelet number, its control and relation to yield per ear in wheat. *Aust. J. Biol. Sci.* **23**, 1–15.

Rawson, H.M., and Evans, L.T. (1971) The contribution of stem reserves to grain development in a range of wheat cultivars of different height. *Aust. J. Agric. Res.* **22**, 851–863.

Shimshi, D., and Ephrat, J. (1975) Stomatal behaviour of wheat cultivars in relation to their transpiration, photosynthesis and yield. *Agron. J.* **67**, 326–331.

Silvey, V. (1978) The contribution of new varieties to increasing cereal yield in England and Wales. *J. Natl. Inst. Agric. Bot.* **14**, 367–384.

Smith, H.C. (1971) Developments in agronomy. Proc. First Ann. Conf. Agron. Soc. New Zealand. pp. 1–8.

Sofield, I., Evans, L.T., Cook, M.G., and Wardlaw, I.F. (1977a) Factors influencing the rate and duration of grain filling in wheat. *Aust. J. Plant Physiol.* **3**, 785–797.

Sofield, I., Wardlaw, I.F., Evans, L.T., and Zee, S.Y. (1977b) Nitrogen, phosphorus and water contents during grain development and maturation in wheat. *Aust. J. Plant Physiol.* **4**, 799–810.

Spiertz, J.H.J. (1974) Grain growth and distribution of dry matter in the wheat plant as influenced by temperature, light energy and ear size. *Neth. J. Agric. Sci.* **22**, 207–220.

Stanhill, G. (1976) Trends and deviations in the yield of the English wheat crop during the last 750 years. *Agro-Ecosystems* **3**, 1–10.

Stoy, V. (1965) Photosynthesis, respiration, and carbohydrate accumulation in spring wheat in relation to yield. *Physiol. Plant. Suppl.* **IV**, 1–125.

Thomas, S.M., Thorne, G.N., and Pearman, I. (1978) Effect of nitrogen on growth, yield and photorespiratory activity in spring wheat. *Ann. Bot. (Lond.)* **42**, 827–837.

Thorne, G.N., Thomas, S.M., and Pearman, I. (1979) Effects of nitrogen nutrition on physiological factors that control the yield of carbohydrate in the grain. In 'Crop Physiology and Cereal Breeding'. (Eds. J.H.J. Spiertz and Th. Kramer.) Pudoc, Wageningen. pp. 90–95.

Waggoner, P.E. (1979) Variability of annual wheat yields since 1909 and among nations. *Agric. Meteorol.* **20**, 41–45.

Wardlaw, I.F. (1970) The early stages of grain development in wheat: response to light and temperature in a single variety. *Aust. J. Biol. Sci.* **23**, 765—774.

Warrington, I.J., Dunstone, R.L., and Green, L.M. (1977) Temperature effects at three development stages on the yield of the wheat ear. *Aust. J. Agric. Res.* **28**, 11—27.

Watson, D.J., Thorne, G.N. and French, S.A.W. (1963) Analysis of growth and yield of winter and spring wheats. *Ann. Bot. N.S.* **27**, 1—22.

Willey, R.W., and Holliday, R. (1971) Plant population, shading and thinning studies in wheat. *J. Agric. Sci.* **77**, 453—461.

Williams, R.F. (1955) Redistribution of mineral elements during development. *Annu. Rev. Plant Physiol.* **6**, 25—42.

Zelenskii, M.I., Mogileva, G., Shitova, I., and Fattakhova, F. (1978) Hill reaction of chloroplasts from some species, varieties and cultivars of wheat. *Photosynthetica* **12**, 428—435.

Zohary, D. (1969) The progenitors of wheat and barley in relation to domestication and agricultural dispersal in the Old World. In 'The Domestication and Exploitation of Plants and Animals'. (Eds. P.J. Ucko and G.W. Dimbleby.) Duckworth, London. pp. 47—65.

14 COMPETITIVE PLANTS, COMMUNAL PLANTS, AND YIELD IN WHEAT CROPS

C.M. Donald

Introduction

Stebbins (1974) has emphasised the need in evolutionary studies to recognise biotic communities that are favourable for the emergence of new adaptive complexes, naming them 'evolutionary cradles'. Though Stebbins was writing of plant evolution above the species level, the concept of an evolutionary cradle at the sub-specific level achieved notable expression during the domestication of crops. The environment for any plant within a crop was so different from that in the wild state, and the selection pressures associated with man's cultural practices so powerful, that strong evolutionary responses inevitably followed. The planting of seed crops, wrote Harlan (1975), set up a whole syndrome of interlocking selection pressures.

The characters favoured by natural selection among wheat plants growing in the wild were prolificacy (the ability to produce many viable seeds), resistance to grazing pressure (particularly through a prostrate habit of growth), wide seed dispersal, and strong germination and establishment. In the cultivated crop the principal criteria of success, in addition to prolificacy, were an exacting requirement to compete against weeds and other wheat plants in the absence of defoliation by animals, and successful passage of the seed from the ripe ear, through the artificial processes of harvesting, threshing and storage to the seedbed in the following season. Natural seed dispersal and resistance to grazing were no longer significant. But, as discussed below, the criteria leading to success in natural selection, and those leading to esteem by man, had different and sometimes opposing features.

The selection processes within wheat crops from the beginning of cultivation until the advent of modern plant breeding have been extremely diverse in manner and kind (Table 14.1). 'Conscious selection by man' and 'natural selection', two of Darwin's categories, provide obvious groupings among these processes. Within natural selection, a further subdivision is shown, using terms proposed by Nicholson (1962), namely 'environmental selection' processes and 'selection through competition'; these aspects of natural selection were recognised by Darwin but not sharply differentiated.

Table 14.1

The components of selection in wheat crops from about the beginning of wheat culture until the advent of plant breeding.

A – CONSCIOUS SELECTION BY MAN: for larger ears; larger or plumper grain; higher yield of grain per plant; for many qualitative characters e.g. grain colour, shape, flavour etc.

B – NATURAL SELECTION: of those plants with the greatest number of viable seeds produced in the crop and carried forward into the grain sown in the following season.

I. **Environmental Selection** *of those plants best fitted to the regional and cultural environment, almost independently of their relationship to other plants within the crop community.*

   (i) Natural Selection for Adaptation to the Regional Environment :
- to the climate: to the length of day, the length of the frost-free season or the effective rainfall season, etc.
- to the soil type: to the soil texture, pH, nutrient status, salinity, etc.

  (ii) Natural Selection for Adaptation to Man's Cultural Methods :
- to the sowing rate; for non-dormancy at sowing time; establishment from variable depth of planting, associated with primitive sowing methods; synchonous ripening of all ears of the plant and grains of the ear; tough rachis at harvest time; brittle glumes, permitting easy threshing; and loss of seed dispersal and burial mechanisms (sharp-pointed grain, awns, fragile rachis, etc.).

II. **Selection through Competition** *of those plants capable of survival and prolificacy in crops of mixed genotypes or in weedy crops.*

   (i) Natural Selection for : prolificacy; seed size; and rapid germination.

  (ii) Natural Selection for : height; a canopy of large horizontal leaves; free tillering habit; a widely ramifying root system; and a reduced rate of photosynthesis per unit leaf area. (The features in this subgroup are considered to be potentially harmful to yields in pure stands).

Selection during domestication

Selection processes in wheat crops

Environmental selection occurs when some plants within the population show better adaptation than others to local climatic, soil or cultural conditions. It may proceed almost independently of any competition between plants, as when early flowering individuals in a crop are prevented from seed setting by a late frost, or when late flowering plants produce no grain because of hot, dry conditions. A composite barley population, involving crosses between 31 cultivars (Allard and Jain, 1962) showed rapid adaptation to the climate at Davis, California. The population shifted strongly to earliness in heading date for 5 generations and then more slowly for the next 15, a directional selection. At the same time there was a steady elimination of 'the tails' (either very early or very late), so that variance decreased, representing a stabilizing selection around the optimum heading time for the locality. Similar environmental selection has been recorded for other climatic features, especially length of day, and for adaptation to soil features, such as texture, pH, fertility and salinity. "All those who have closely attended the subject insist on the close adaptation of numerous varieties (of wheat) to various soils and climates even within the same country" (Darwin, 1868).

Strong environmental selection occurred for adaptation to man's cultural methods (Harlan *et al.*, 1973); it included natural selection for a tough rachis at harvest time, for synchronous ripening of the ears and for non-dormancy of the grain at the following sowing season. Natural selection for some of these features must have occurred not over millennia but within a few generations. Seed dispersal and burial mechanisms, which had no value in cultivated wheats, tended to disappear.

Although the responses of wheat crops to regional and cultural conditions have been widely recognized, there has not been as much emphasis until recently on 'natural selection through competition', though this is the classical expression of Darwinian selection. It is with selection through competition and its outcome in production and plant breeding, that this paper is particularly concerned. Within a wheat crop of environmentally-adapted but diverse genotypes, the most competitive and successful plants have distinctive features: tallness; a leafy canopy; free tillering; the prolific production of seeds of adequate size; rapidly germinating seeds; and an extensive root system.

Selection for height

Of all the features giving competitive advantage within either natural plant communities or crops, height is paramount. Throughout wheat culture, there has been strong directional selection towards increased height, due to the advantage of tall plants in the interception of light. Jensen and Federer (1964) showed that when a tall wheat variety (Genesee, mean height 122 cm) was grown in rows alongside shorter wheats (mean height 112 cm), its yield was enhanced by 9% relative to its pure stand, while that of the shorter wheats was depressed by 5%. And when a hybrid bulk population from a tall x short wheat cross was carried from F_2 to F_6 (Khalifa and Qualset, 1975), the proportion of short plants fell from 21.9% in the F_3 (21.0% predicted on genetical criteria) to 9.4% in the F_6 (compared to 24.9% predicted). Similar data are available for other cereals. But as well as directional selection for tallness, there has been stabilising selection. Very tall plants suffer exposure and shattering of ears, and, perhaps more importantly over millennia, tallness must have been associated with a smaller proportion of assimilates passing to the grain, expressed as a falling harvest index (Donald and Hamblin, 1976) and reduced prolificacy and yield. In addition tall crops lodge more readily, and make harvesting difficult.

All these factors influencing selection for height brought about stabilising selection to give wheats which by today's standards were 'very tall'. Percival (1921) described *Triticum turgidum* as up to 180 cm in height, *T. compactum* to 140 cm and *T. aestivum* to 150 cm and there can be no doubt that these tall varieties were developed by natural selection for height. Tall cultivars were often preferred by farmers because of the value of their straw for bedding, thatching etc., but it is believed that these preferred tall cultivars arose by sustained natural selection, not through selection by man.

Selection for leaf characters and photosynthesis

Natural selection in favour of wide, horizontal or floppy leaves must also have occurred throughout the culture of wheat. Such plants have strong competitive advantage through preferential interception of light. The influence of canopy structure on competitive ability was demonstrated by the comparative behaviour of wheat varieties of different leaf habit (Tanner *et al.*, 1966). In a weedy situation, the floppy leaved varieties were able to suppress weeds and yield well, while those of erect leaf habit were depressed in yield. In a weed-free situation the yields were reversed. The advantage of large leaves as 'competitive weapons' was also shown in the study of a barley F_3 population at a strongly competitive spacing (Hamblin and Donald, 1974), where there was a positive correlation between leaf length and grain yield. A stabilizing factor in the development of leafiness in wheat in some environments may be the tendency for large, floppy leaves to be associated with more prolonged growth and lateness in maturity or with a reduced harvest index.

The plant breeder has not looked perceptively at the foliage of his wheat plants but only at the ear and grain. Darwin (1868) remarked that "if a (wheat) plant with peculiar leaves appeared, it would be neglected unless the grains of corn were at the same time superior in quality or size". This attitude has continued among most wheat breeders until recently.

A surprising feature of the evolution of wheat under domestication is the much lower photosynthetic rate per unit area of leaf of modern wheats compared to primitive *Triticum* and *Aegilops* species (Evans and Dunstone, 1970; Evans and Wardlaw, 1976). The authors suggested that this falling rate might be due to the reduced surface/volume ratio of the mesophyll cells, which Kranz (1966) had shown to have become progressively larger during domestication. However, Evans and Dunstone found that leaf area had increased more than photosynthetic rate had fallen, so that photosynthesis per leaf was much greater in modern wheats. They also noted a positive relationship between leaf size and grain size, and reasoned that selection for yield would tend to lead progressively to increased grain size, increased leaf size, larger cell size and lower photosynthetic rate.

An alternative explanation of the enigma of falling photosynthetic rates during the evolution of wheats under domestication is based on competitive relationships. Within the crop canopy, plants with large leaves, usually wide, floppy, drooping leaves, have had strong competitive ability for light and clear selective advantage throughout domestication. As long as the photosynthetic rate per leaf was sufficiently maintained, a progressive increase in leaf size ensured natural selection, with a consequent relaxation of selection pressure on the photosynthetic rate per unit of leaf area. Less leafy plants were at great selective disadvantage due to shading by their leafy neighbours, which, though they might be 'physiologically weaker' were 'ecologically powerful'. Thus there would again be stabilising selection − between directional selection for larger leaves in the competition for light, and opposing directional selection for an adequate or advantageous rate of photosynthesis. It is proposed that the modern wheats have leaf sizes and photosynthetic rates representing the outcome of that stabilising natural selection. Selection for yield, it is suggested, has not been involved as the initiating factor, and man has played a role in the falling rates of photosynthesis only through his crop production.

Selection for tillering

The selective advantage of free tillering is illustrated by the response of wheat to poor establishment, which was doubtless common in early agriculture. When the density of a wheat stand was reduced from 184 to 35 plants m^{-2}, the number of tillers per plant increased from 5.5 to 13.7, and the number of grains per plant from 46 to 215, with no significant change in yield per square metre (Puckridge and Donald, 1967). Similarly, Bremner (1969) found that halving the density of a wheat crop from 303 plants m^{-2} increased the contribution of grain by the tillers from 30% to over 50% with a yield reduction of only 9%. Each of these studies was with a genetically uniform cultivar. It is clear that if a genetically diverse crop were depleted in plant numbers for any reason, free tillering genotypes would have great selective advantage over less tillered kinds. Under domestication there seems to have been little reduction in tillering capacity (though there has been with increase in ploidy, Evans and Dunstone, 1970), and modern cultivars are capable of producing 30 or more fertile tillers (ears) per plant at wide spacing, though only two or three when competing at crop densities (Puckridge and Donald, 1967). Free tillering has been maintained during domestication and until plant breeding began.

Selection for seed size

The increase in seed size from the earliest wheats to modern wheats, a modest doubling from the wild *T. thaoudar* (up to 20 mg) to *T. aestivum* (40-45 mg), is commonly regarded as a tribute to man's ability as an agriculturalist. But changes in seed size are of great complexity, with the influence of natural selection probably greatly exceeding that of selection by man.

The smaller size of the seed of wild plants is presumably because such plants had a selective advantage. They were likely to have had more seeds than large-seeded plants, and their small seeds were more easily dispersed and buried.

Under cultivation (considered at first without selection for seed size by man) the advantages of small seed in dispersal and burial were no longer relevant. Christian and Gray (1941) found that when large seeds (45 mg) were sown alternately with small seeds (27 mg) of the same variety within each row of a wheat crop, the large seeds gave large seedlings, leading to plants which out-yielded those from the small seeds by 57 percent.

This greater productivity by plants with larger seeds would suggest strong directional selection within early crops towards greater seed size. But prolificacy was involved as an opposing factor (Stebbins, 1974; Harlan, 1975).

In Christian and Gray's study, seeds of the two sizes were of the same genotype and predictably they produced grain of the same size. But if the difference in the size of the seed had had a genetic basis, so that the 45 mg seed in the mixture had yielded grain of 45 mg, and the 27 mg seeds grain of 27 mg then, although the large seeds gave 57% greater yield, they would have produced 6% fewer grains.

Thus two factors, the greater competitive advantage of plants with large seeds and the greater prolificacy of plants with small seeds would eventually lead to stabilising selection for seed size within wheat crops of mixed genotypes.

Early in the history of cropping, a further additional factor came into operation — the conscious selection by man of larger and plumper grains, because these were, and still are, associated in the minds of growers with high yield. Several Greek and Roman writers emphasized the importance of retaining large grain from the harvest, to be sown the following year (Percival, 1921). Large grains could be separated during winnowing or by skilfully shaking grain on a shallow tray. Thus an added and powerful selective advantage, unrelated to field performance, lay with plants producing larger seeds. Though continuing recombination and segregation would still have ensured stabilising selection through small-seeded prolific genotypes, the equilibrium size would presumably have tended to increase.

But perhaps a factor reducing man's concern for the selection of large seed was that it may have added little to yield. Continuing with the example of Christian and Gray's study, there was no difference in the yield by crops established wholly from large seed (45 mg) and those wholly from small seed (27 mg) of the same genotype. And where genetic selection has been made for large seed size, there has usually been a compensating decrease in the number of grains per plant (Grafius *et al.*, 1976). Indeed these authors' data indicated that the best means of increasing yield may be to select for grains per head and to allow seed size to move as a more or less random variable.

It may be reasonable to conclude that though man has taken a strong interest in larger wheat seed, the influence of natural selection has been so all-pervading and continuous, that his influence, at least until plant breeding began, may have been quite limited.

Selection for speed of germination

In the wild, the irregular or protracted germination of wheat or other annual grasses, is a partial protection against uncertain climatic conditions — so called 'false starts' to the rainfall season. But cultivated wheats germinate more rapidly and evenly than wild wheats (Evans and Dunstone, 1970).

This evolution to speedy germination was due wholly to natural selection. Within a crop sown in a prepared seed bed, plants that emerged first had strong competitive advantage over their neighbours. For example, within a uniform barley cultivar, sown as a drilled crop at a normal field rate, plants that emerged one day earlier than their neighbours had seedlings 16% heavier at seventeen days from sowing, and yielded 14% more grain (Soetono and Donald, 1980). There was doubtless strong directional selection in early crops towards simultaneous and rapid germination by all seeds, a feature of modern cultivars.

Selection for root characters

Knowledge of the role of root systems in natural selection under domestication is seriously lacking. If it is reasonable to extend to the root system our understanding of features of the plant tops giving natural selection through competition, one would suggest that just as a tall, leafy, tillered plant secures an undue share of the light, so would plants with a widely ramifying root system have selective advantage, by absorbing water and nutrients more rapidly and more extensively than plants with more restricted roots. The implications of different kinds of roots are discussed later.

Competition and crop yields

Darwin emphasised that "natural selection, it should not be forgotten, can act solely through and for the advantage of each being". One can apply this thought to a wheat crop by recognising that natural selection is not towards the greater growth or productivity of the crop, but only a response by the community to the greater fitness of particular individuals. In the case of plants in a wheat crop, Darwinian fitness is the capacity of any plant to produce more seeds and seedlings than its neighbours. But there is no corollary that the total yield of the crop in which those successful plants are growing will be increased. All that follows is that the proportion of plants of the successful genotype will be greater in the ensuing generation.

More importantly, there can be no assumption that if the successful genotype were to increase over succeeding generations to become the sole genotype in the crop, there would be any enhancement of crop yield. "Natural selection is solely for the advantage of each being..." and that being is the plant, not the crop (nor man).

"Every gene combination that can survive in competition with all others is *de facto* superior", wrote Mayr (1949). In the evolutionary context in which those words were written, this is a truism. In the agronomic sense of capacity to yield more grain as a crop, competitive success promises nothing. With this viewpoint in mind it is of interest to examine which of the features favoured by natural selection in wheat crops of mixed genotypes may favour or depress crop yields in the modern situation of a pure stand.

It seems evident that all forms of environmental selection for adaptation to the regional climate or soil, or to man's cultural methods, must be positive in their influence on yield, not only of the selected plants but also of the whole community. The crop will grow better and produce more if it has evolved through plants progressively selected for adaptation to the limits of the growing season or the length of day or the soil texture or salinity. Similarly, there can only be gain in the effectiveness of establishment if all the seed is non-dormant when sown, and in the harvested yield if all rachises are tough and all grains ripen simultaneously. Provided the physical environment and cultural methods are constant, only benefit can accrue from environmental adaptation. But if the environment changes in the course of crop production, major adaptive change may be needed. This was illustrated by the concurrent adoption of heavy nitrogen application and the development of semi-dwarfs resistant to lodging (Vogel *et al.*, 1963).

When we turn to natural selection governed by competition, the situation is very different from that for environmental adaptation. Some features for which natural selection has occurred through competition will contribute to increased grain yields in pure wheat stands, but it is proposed that others are undesirable features in crop production. It is emphasised that this distinction (shown in Table 14.1) is not based on differences of evolutionary significance. All natural selection due to competition favours effective competitors − it is beneficial to those plants. The distinction into two groups is based only on *post hoc* criteria − an agronomic assessment by man.

Prolificacy, the production of many seeds by the individual plant, must have been a principal factor in the choice of wild material for domestication and in any early advance in crop yields. But during crop evolution, prolificacy was significant only in relation to the performance of other plants in the crop. It was not synonymous with the man-made concept of yield, expressed as amount of grain per unit land area. However, since prolificacy is a direct component of yield (the product of prolificacy and grain size), it might conceivably lead to a direct relationship between competitive success during crop evolution and high performance in pure stands. Such however, is frequently not the case.

Yield in mixtures and pure stands

Many instances have been reported of an inverse relationship between yield in competitive situations, as during domestication, and yield in pure stands. At Davis, when two cultivars were equally mixed and then repeatedly re-sown, the most productive and desirable genotype (25% superior yield in pure stands) was reduced to 22% of the plant population after three years (Khalifa and Qualset, 1974). In Hamblin and Donald's (1974) study of barley, the plants with the highest grain yield under intergenotypic competition in the F_3, gave rise to the lowest yielding lines when tested in plots in the F_5. And Wiebe *et al.* (1963) concluded that "when high yield is the criterion selected for, say in the F_6, and the selection is intended for use in pure stands, the instructions from the present study are that one should save the poorest plants from the F_6 rather than the good ones". But in some instances a positive relationship has been recorded. Jensen and Federer (1965) found the same ranking in four wheat varieties in pure stands and in mixtures with each other. And Johnston (1972) recorded a correlation of 0.55 between the yields of 100 homozygous barley varieties in pure stands and as equal components of a single mixture in the same season.

The situation closest to that in early wheat crops is in bulk populations and varietal mixtures which have been continually harvested and re-sown for many years without selection. Suneson (1949) sowed an equal mixture of four barley cultivars; after 16 years the highest producing variety in pure stand had been reduced to 0.4% of the mixture, a striking example of contrasting performance in competition and in pure crops. In the well-known study by Harlan and Martini (1938), eleven barley cultivars were planted in equal mixture at 10 stations in the United States. At all centres, cultivars of known agricultural inferiority were rapidly eliminated. The most successful component at most centres was the leading local cultivar, but at several of the stations it was another cultivar.

It is suggested that all these results with mixtures, bulk populations and pure stands, some of which appear inconsistent, are embraced within the following relationships:

(a) If locally adapted and poorly adapted genotypes are compared in mixtures and pure cultures, there is a possibility and even a likelihood that they will rank in a similar order in both situations. This was a component of the results of Harlan and Martini, and of Johnston.

(b) If the comparison is for only a single season and if there are many lines in the mixture, then a low yielding genotype of strong competitive ability may make only a modest gain in its share of the environment. Yet over several seasons it may be capable of attaining dominance (Allard *et al.* 1968).

(c) The experimental situation of particular significance in relation to domestication is the behaviour of a mixture of genotypes all of which are adapted to the regional environment. If relatively high yielding genotypes in mixtures give a relatively low yield in pure stands, then their performance in mixtures can be due only to their strong competitive ability against the other genotypes.

Many of the foregoing studies fall into this third category and it remains to ask what characters lead to converse performance in mixtures and pure stands.

Features involved in converse performance

Tallness and leafiness have been shown to give strong advantage in mixtures, but a relatively weaker performance in pure stands e.g. in Hamblin and Donald's study (tallness and long leaves) and in Khalifa and Qualset's experiments (tallness and competition for light). The influence of tallness as a factor of competitive advantage and its reversal in pure stands was analysed in detail by Fischer

(1978). Yecora 70 (85 cm in height) yielded 10% more than Jupateco (110 cm) in pure stands. But when the two cultivars were grown in alternate 4-row plots (whole plot harvested) the tall Jupateco had a 2% greater yield, increasing to 26% in 2-row plots and an estimated 60% advantage in alternating 1-row plots. In spaced plantings of many lines (40 cm x 40 cm), the increase in yield of taller genotypes was 0.58% per centimeter height superiority over neighbours.

Data on the influence of 'leafiness' and leaf disposition on the productivity of wheat and barley are less abundant, because these features are less easily measured than height. A striking relationship was recorded by Tanner *et al.* (1966) who ranked some 300 varieties of wheat, barley and oats for leaf width and leaf angle. They correctly categorised all but two of the 50 highest yielding strains purely on the narrowness and erectness of their foliage, relative to the wideness and floppiness of the leaves of the other varieties, and advocated this kind of visual evaluation as a criterion for yield in Ontario. Hamblin and Donald (1974) similarly found a negative relationship in F_5 field plots between leaf length and yield among the competing plants of the F_3 generation. In his studies in Mexico, Fischer (1978) compared the yields in 3-row plots, in which the plants suffered less competition because each of the outer rows was bordered by an "unsown row", with yields by the same wheat varieites in true crop situations. Erect leaved types showed relative advantage in the highly competitive crop situation, those with more horizontal leaves, in the 3-row plots. All these results indicate the value in crops of narrow erect leaves in contrast to the long, floppy leaves that are advantageous in mixtures.

A third proposition — that free tillering, while increasing competitive ability, detracts from yield in pure stands — is a controversial issue. There is a traditional belief in the virtues of all yield components — that the best wheat plants will have more ears and/or more grains per ear and/or larger grains. But many studies have shown negative relationships among these features, so that increasing any one of them is likely to contribute little to yield. Evans and Wardlaw (1976) propose that these negative relationships and the extreme plasticity of the yield components in wheat, including tillering, give 'many paths to success as a crop' and confer an extremely important capacity to adjust to environmental conditions as growth proceeds.

The greater yields of non-tillered (uniculm) wheat and barley plants compared with tillered cultivars is discussed later. Islam and Sedgley (1980) have pointed out, however, that the uniculm and tillered genotypes used in those experiments were not isogenic lines. They therefore tested the effect on yield of restriction of tillering, by surgery, to two culms per plant in a wheat crop in a low rainfall cereal environment (Merredin, Western Australia; with 180 mm of rain in the growing season in each of two test years). With similar biological yields, the de-tillered plants had a considerably higher grain yield than the controls, 1.94 compared with 1.65 t/ha (19% greater in 1978 and 22% in 1979). Islam and Sedgley further demonstrated that the greater yield of the de-tillered crop was associated with a lesser use of water early in the season, leaving more water for use after anthesis. Thus there are clearly situations in which free tillering is disadvantageous in pure stands.

Large seeds and rapid germination, of such significance in competition during crop evolution, are not of value for those reasons in a pure stand. There is no evidence that leaf size in the developed crop canopy or ultimate dry matter or grain yield will be influenced by mean seed size. But large seeds and rapid germination are features valued for other reasons in wheat production. They enable vigorous establishment, especially by giving larger seedlings, and more rapid early progress in leaf area index, though those features, *per se*, do not necessarily mean greater yield. A further advantage of a capacity to produce large grains is that it adds to sink size and thus to potential grain yield in modern crops.

In Suneson's study with four barley varieties in a repeatedly re-sown mixture, root systems were invoked as the factor of competitive advantage. The successful cultivar, Atlas, compared to the suppressed high yielding cultivar, Vaughn, had a strong, more freely branched, though shallower root system (Lee, 1960). This apparently gave Atlas an advantage in securing water and nutrients from the surface soil. But in pure stands the root system of Atlas gave no advantage and Vaughn had greater yield.

The communal plant

Broadly it may be said that prolificacy, seed size and speedy germination are features which, having been naturally selected through competition, are also of value for grain yield in pure stands, though not necessarily for the same reasons. On the other hand the evidence suggests that wheat and barley plants that were naturally selected during domestication because of competitive ability due to tallness, leafiness, free tillering or free branching root system are unlikely to prove the most productive in pure stands. Theoretical considerations to support this view were summarised (Donald, 1968) as follows: "In a field crop ... the individual plant within the community will express its potential for yield most fully if it suffers minimum interference from its neighbours. Thus its neighbours should be weak competitors. And since, for the purpose of this discussion, all plants in the crop are of like genotype, then the ideotype itself [a plant model fitted to any specified environment] must be of low competitive ability... While strong competitive ability is advantageous against other species such as weeds, it will lead in a monoculture to intensified competition and heavy mutual depression among the crowded plants".

The term 'communal plant' is proposed, defined as follows:

A communal plant has features in accord with the success of the crop community rather than of the plant itself. It is adapted to existence within a pure stand; it is of weak competitive ability, so that it interferes with like neighbours to a minimum degree; in mixtures with other genotypes it is likely to be suppressed; communal plants may give low individual plant yields, but, when grown in a pure stand at a density sufficient to induce interplant competition and full exploitation of the environment, they are capable of high crop yields.

It is proposed that any ideotype for wheat or barley crops should be based on communal plants.

Selection by man

It is hardly possible that early cultivators could have selected plants directly for their potential to give high crop yields. Even today, that presents great difficulty. The characteristic cereal crop until a century ago was, and in many regions still is, a land race — a highly integrated mixture of genotypes adapted to the environment, the cultural practices and the use to be made of the grain (Harlan, 1975). The maintenance of land races is itself an indication of the limited selection imposed by man.

Indirect selection for wheat yield through large ears or large grains involved not only partially conflicting criteria of advance, but also had to operate among the intricacies of natural selection forces in a varied and heterozygous land race. As in other crops, man's most active selection may have been for cosmetic or culinary characters — seed colour, ear shape, or cooking quality, without consideration of yield.

We can thus envisage selection in wheat by man as lying across a full spectrum — selection which might have increased yield (as of prolific ears), selection which might have depressed yield (as of competitive plants) and selection which tended to neutrality or, even today, would be of unpredictable effect (as of seed colour). After early advances in yield, an equilibrium may have developed, in which man's selection became little more than a sustained supplement to the ceaseless, all-pervading, and powerful operation of natural selection. Man's great contribution to wheat yields was cultivation and annual resowing — he built and rocked the 'evolutionary cradle', while natural selection guided the evolving crop.

Many writers question the extent of our progress in increasing the yield potential of our wheat cultivars. Darwin (1868), in discussing the value of selection, wrote that 'the little which man has effected, by incessant efforts during thousands of years, in rendering plants more productive.....would seem to speak strongly against its efficacy". Harlan (1975) quotes evidence of substantial wheat yields in the Near East 4000 years ago. Looking at more recent times, Athwal (1971) questions whether there was any real advance, in the absence of disease, in wheat yield potential in U.S.A. during this century, before the advent of the semi-dwarf cultivars.

The selection of the unfit

Since there has been long-continued natural selection of wheat plants with competitive features, and since certain of these features are disadvantageous in pure stands, it is desirable that plant breeders should seek to modify or reverse them. We may ask to what extent this has already occurred during the decades in which wheat and barley breeding has been systematically practised, or to what extent it is now in prospect.

There appear to have been two impediments to the reduction of 'competitive features'; there was hesitancy in accepting that any lesser plant attribute, whether less height, less leaves, less tillers, less ears or less anything, might be advantageous; and secondly, many of the assessment techniques used by breeders did not (and still do not) recognise the powerful interaction between plant spacing, plant type, competition and yield.

Fig. 14.1 shows three definitive dispositions of wheat plants in relation to genetic uniformity and density recognised by Donald and Hamblin (1976). They considered the kinds of wheat plants that would be successful in these three situations, and concluded that: successful *plants in isolation* would be of lax habit, free tillering and with an extensive display of large leaves; successful *plants within a community of mixed genotypes* would be of strong competitive ability, tall, with a canopy of long, wide leaves, able to shade neighbours, and with many tillers; and *wheat plants in a pure stand* (a successful crop) would be of weak competitive ability, dwarf stature, with sparse or nil tillering, and few, small erect leaves.

Figure 14.1 also gives a representational account of various plant and plot arrangements in common use in plant breeding. It is evident that many of these arrangements will tend not to select those kinds of plants likely to be successful in a crop. The assessment of genotypes growing as spaced plants will tend to favour free tillering, floppy leaved, non-erect forms, while as already discussed, within crowded, mixed or segregating communities, tall leafy plants will succeed.

Some reversal of competitive features is occurring, though not based (except in very few programmes) on the philosophy of breeding communal plants. The most striking trend of this kind in wheat is the marked reduction of height. Law *et al.* (1978) refer to a reduction in the height of wheats in Britain between the 1930's and early 1960's of about 30 cm from initial heights in excess of 130 cm.

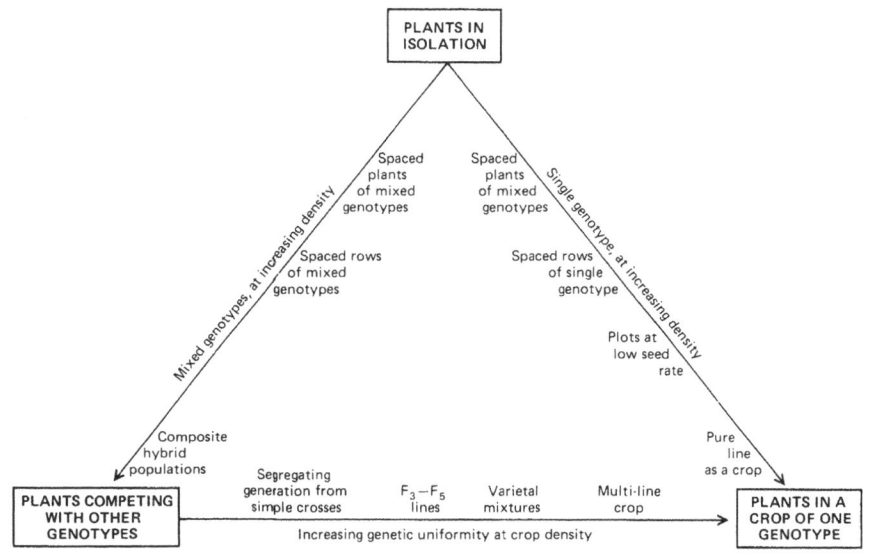

Fig. 14.1. The three definitive dispositions of wheat plants in relation to genetic uniformity and density of planting, together with intermediate arrangements commonly used in plant breeding. (Developed from Donald and Hamblin, 1976).

But the most dramatic height reductions occurred with the use of the dwarf genes of Norin 10, first in Japan and then progressively in Washington State, at CIMMYT, Mexico, and in most wheat areas of the globe. Law *et al.* point to the release of semi-dwarf wheats of 70-80 cm in the later 70's as the second step in height reduction in Britain.

Any trend towards even further reduction in height is viewed by wheat breeders with uncertainty. This is partly because of harvesting problems, but especially due to a widespread belief that within semi-dwarf material, taller lines will give higher yields. Law *et al.* (1978) found a positive relationship between height and yield in semi-dwarf lines, both among spaced plants (30 x 10 cm) and in small, drilled plots. However, Fischer's studies (1978) clearly demonstrated that in such plant and plot arrangements as these, the apparently greater yield by taller plants may be due wholly to their competitive advantage over shorter neighbours. The difference in relative performance of tall or dwarf plants growing under intergenotypic competition and intragenotypic competition respectively is becoming increasingly recognised as a relationship interfering with the assessment of still shorter material.

The move away from a competitive 'shading' canopy towards one comprising shorter, narrower, more erect or less abundant leaves is not so well documented as changes in height but is nevertheless a clear trend in wheat and barley breeding. Vogel et al. (1963) attributed the success of their semi-dwarf wheats in east Washington as being due in part to the medium to small length and width of the leaves. Harlan (1976) refers to the development of shorter and less vegetative barleys, MacKey (1978) to the increased tendency to breed for a wheat plant with a less luxuriant vegetative growth, and CIMMYT (1979) to the good performance in narrow rows of short wheats of compact habit and with very erect leaves. Similarly Donald (1962, 1968) advocated less leafy canopies and more erect leaves in both forage grasses and cereals. These viewpoints are a challenge to those wheat breeders who have regarded leaf characters as irrelevant to breeding for increased grain production.

The low photosynthetic rate per unit leaf area in modern wheats may seem a potential limitation to future productivity, though not necessarily of the productivity of existing cultivars. However, if the thesis submitted earlier is in fact valid — that there has been stabilising selection between effective competition for light and an effective photosynthetic rate — then there seems opportunity for substantial gains in yield. This requires on the one hand small leaves with small cells and high photosynthetic rate, and on the other hand a controlled and adequate leaf area index through the dense planting of uniculms and a canopy of good light profile through leaf erectness.

The breeding of wheat with a reduced number of tillers began earlier than the move to less leafiness. MacKey (1966) records a trend in wheat varieties in N.W. Europe towards decreased tillering with larger heads and larger seeds, and more recently (MacKey, 1978) reports a reduction to 1.2 ears per plant in densely sown wheat crops in Sweden, involving agronomic as well as genotypic changes. However Vogel et al. (1963) considered high tillering capacity to be an important attribute of their semi-dwarf wheats. Jones and Kirby (1977) advocate cultivars with a small number of tillers, as giving sufficient flexibility, while effectively reducing the harmful competition between excess tillers, as they develop, and the growing spikelet primordia of the main stem and early tillers (Kirby and Jones, 1977). The extreme viewpoint is that the ideal wheat plant has a main stem and no tillers, as discussed later.

Thus there are trends, some of them ill-defined or erratic, away from many competitive characters that have developed over centuries by natural selection. There is need however for more positive recognition by breeders that these consequences of natural selection must be reversed and that testing techniques should not favour competitive plants.

Natural selection, in Darwinian terms, depends on the fitness of the plant in the community and the environment. But the breeding of wheats that are productive as crops will depend, in respect of these competitive features, on the selection of the unfit — on the selection of communal plants of weak competitive ability within the environment in which they are to be grown.

Breeding programmes based on a communal ideotype

There are currently two principal philosophies in plant breeding — 'defect elimination' and 'selection for yield'; a third approach has been proposed, namely the design and breeding of ideotypes, model plants for defined environments and uses (Donald, 1968).

Wheat breeding based on selection for yield has certain elements in common with natural selection. Variation is created, much as in nature, though with substantial skill and judgement. Thereafter selection of the fittest is undertaken, not in terms of ability to compete and reproduce in mixtures, but in terms of grain production by a pure stand.

On the other hand, the design of ideotypes is almost diametrically opposed to natural selection. Because natural selection follows the success of individual plants within a community in a current environment, there can be no anticipation, as Darwin emphasizes, of any future environmental situation. But if a successful crop ideotype is to be designed, the environment, including any future modification of cultural practices must be assessed or conceived. The ideotype and the environment, to the extent that each is flexible, must be effectively integrated with one another in order to reach maximum crop production.

A wheat ideotype for high grain production was designed (Donald, 1968), based on the following premise — that the environment would be non-limiting for water or nutrients; that the plants would be weak competitors; that they would have a favourable movement of assimilates to the ear (a high harvest index); and that the cultural environment would be a crop of high density sown in narrow rows.

The ideotype had a short, strong stem, few small erect leaves, many grains per unit weight of biological yield (a high harvest index), and erect ear, awns, and a single culm — that is to say, it depicted a communal plant of high harvest index. It is of interest to consider whether it can now be usefully modified and whether its relationship to environmental conditions needs review.

Atsmon and Jacobs (1977) in Israel developed lines of wheat which they described as 'quite similar' to the Donald ideotype. These lines had restricted tillering (some were uniculms; some, with few tillers, were named 'oligoculms') and gigas characters (thick, stiff straw; broad, thick leaves; and large spikes with many spikelets, florets and grains). These uniculms and oligoculms had been developed by crossing uniculm selections from a North African local cultivar with genetic sources of dwarfness, semi-dwarfness and disease resistance. Under field conditions, a gigas uniculm gave a grain yield per plant, and per metre^2, 13% greater than that of a standard, tillered cultivar, Hazera 2152.

In South Australia, a breeding programme was undertaken to develop a barley line to accord with the 1968 ideotype. The basis of the barley programme was the incorporation into an appropriate background of two recessive genes, namely the non-tillering gene from Kindred Uniculm barley and the dwarfing *uzu* gene from a semi-brachytic Japanese cultivar, Suifu. The non-tillering gene alone gave very tall uniculms prone to severe lodging, but the *uzu* gene reduced them to about the same height as the standard South Australian barley cultivar, Clipper. Lines developed in this programme were early maturing, uniculm, six-row barleys, moderately resistant to lodging. However they had lax leaves of medium length in contrast to the short erect leaves of the ideotype. They were selected for height, harvest index and less leafy habit, without reference to grain yield or quality. When tested in field plots, from 1974 onwards, these lines gave yields (depending on season and seed rates, see below) of 100-135 percent of tillered lines, commonly in the range 115-125 percent (Donald, 1979).

Some features of the lines based on a communal ideotype

The Israeli uniculm wheats and the Australian uniculm barleys have many features in common - a heavy stem, large leaves, a high number of grains per ear, early maturity and a high harvest index.

Major pleiotropic effects of the uniculm genes seem to follow from the availability of assimilates otherwise passing to developing tillers. Kirby and Jones (1977), conclude from their study of the surgical removal of tillers from barley plants, with a consequent increase in the number and weight of grains in the main stem, that "during their early development, tiller buds are competing strongly for a limited supply of resources". The same explanation is offered for the tallness (unless a dwarfing gene is introduced), thick stem, large leaves and greater root growth of genetic uniculms (Donald, 1979). Stem, leaf and root meristems are favoured when tiller meristems are absent. Some of these features, e.g. large leaves, are in conflict with the ideotype, which has few, small, erect leaves.

The ecological significance of the considerably greater root development by barley uniculms (66 percent more roots than tillered plants with an equal weight of tops, in Donald's (1979) pot culture study) remains to be determined. It seems likely to be a general feature of uniculm cereals because all culms in the crop will have a seminal root system. Passioura (1972) has shown that a restriction of water uptake by wheat plants in pots may be advantageous if the water supply is limited, so that more water is left in the soil for use during grain filling. This may be attainable through the use of cultivars with fewer seminal roots or with roots with smaller xylem vessels, restricting water flow. But the contrary view that a strongly developed root system may permit the exploitation of deeper water cannot be discarded. At least on some Australian soils, dryland wheat crops may fail to make use of all available water in the root zone (Schulz, 1971; Walter and Barley, 1974). There is urgent need for a fuller study of the structure and function of root systems both in mesic and semi-arid wheatlands.

A notable feature of the tillered and uniculm wheat and barley lines was the considerable difference in the ratios of yield components (Tables 14.2 and 14.3). Tables 14.2 and 14.3 contrast the yield components of tillered and uniculm wheats and barleys. The uniculm wheats, though sown only at the same density as tillered cultivars, gave a greater grain yield per square metre, mainly because of the high number of grains per spikelet, alone almost compensating for the lesser number of ears per plant. In the barley lines, the factors leading to the greater yield of the uniculm were high plant density and again the increased number of grains per ear.

Figure 14.2 shows the relationship of sowing rate, water supply and grain yield at Roseworthy, South Australia. It is clearly advantageous in favourable rainfall areas to use a high sowing rate for uniculm barleys in all seasons and, equally, to use a low rate in dry districts. If the rainfall is erratic from year to year, as is the case in most wheat growing regions of the world, a low rate would again seem to be indicated, giving satisfactory yields (relative to tillered lines) in all seasons rather than the 'feast or famine' yields which would result from high seed rates in all seasons.

A high harvest index was named as an essential feature of the ideotype, in contrast to a high grain yield per plant. Donald specifically selected for high harvest index in his barley programme, but the prevention of excessive tillering by any means, whether by genetic modification or by pruning, seems to ensure an improved harvest index (Table 14.4). In Islam and Sedgley's study (1980), with similar biological yields by tillered and of pruned 2-culm stands of plants, the latter gave greater grain yields due to the absence of sterile tillers, water conservation, and the maintenance of a higher leaf area index after anthesis. This is the first clear demonstration in the field of the waste of resources due to the production of later or sterile tillers in wheat.

Table 14.2

The yield components of a tillered and a uniculm wheat, each at 110 plants.m^{-2} (Atsmon and Jacobs, 1977).

Genotype and tillering behaviour	No. of tillers per plant	No. of ears per plant	No. of spikelets per ear	No. of grains per spikelet	Wt. per grain (mg)	Wt. of grain per ear (g)	Yield per plant (g)	Yield per metre^2 (g)
Tillered (Hazera)	5.2	2.6	18.5	2.3	36.0	1.5	3.9	433
Uniculm (Line 492)	1.1	1.0	20.8	5.1	41.4	4.4	4.4	488

Table 14.3

The yield components of a tillered barley at its optimum rate of sowing (75 kg/ha) and a uniculm barley sown each year at 112 kg/ha. (Mean of 1975–6–7, one favourable season and two adverse seasons for rainfall). For other rates see Donald (1979).

Genotype and tillering behaviour	Plants per metre^2	Tillers per plant[†]	Ears per plant	Grains per ear	Wt. per grain (mg)	Wt. of grain per ear (g)	Yield per plant (g)	Yield per metre^2 (g)
Tillered (Clipper)	199	3.78	2.57	12.5	41.0	0.51	1.32	261
Uniculm (WID-101)	309	1.03	0.97	30.2	32.6	0.98	0.95	295

† 1977 only

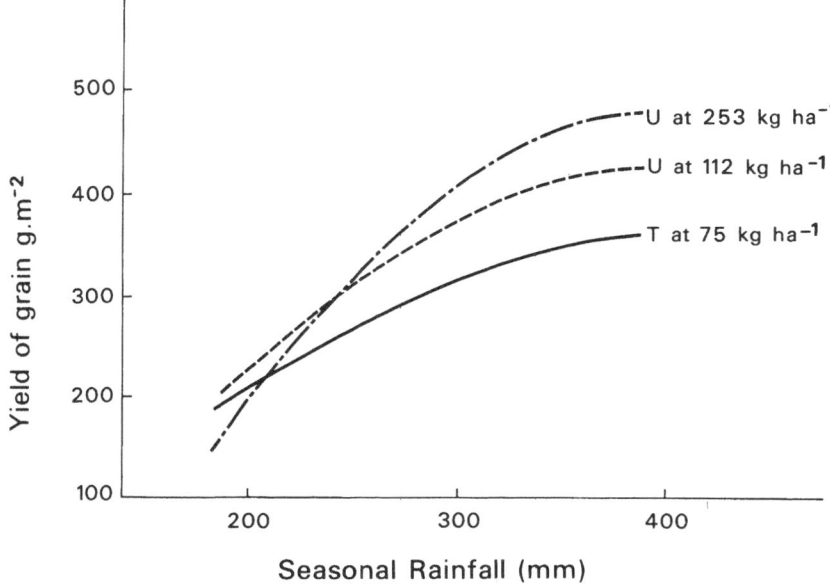

Fig. 14.2. The interaction between amount of seasonal rainfall, sowing rate and yield of a tillered barley cultivar (T = cv. Clipper) sown at its optimum rate (75 kg ha^{-1}) and of a uniculm line (WID-101) sown at 112 and 253 kg ha^{-1} at Roseworthy, South Australia (Donald, 1979).

The control of ear number m^{-2} in crops of uniculms has proved partially successful. Reduced numbers of ears can be realised through lower sowing rates of uniculms, and in less favourable environments this holds great promise of increased yields. In favourable environments, however, the maximum ear density attainable with uniculms has so far not exceeded that with tillered cultivars. But this does not preclude significantly increased ear populations in future uniculm lines with more erect leaves, and better adaptation to high density through improved light relationships within the canopy.

It is perhaps significant that the two principal genes used to develop barley plants approaching this ideotype, were simple recessive genes (non-tillering and dwarfness). De Beer (1958) wrote: "a gene that is non-recessive has gradually become recessive because there has been a selection of gene complexes in favour of those which suppress of effect of this gene". It is suggested that many potentially valuable genes have been suppressed during the domestication of wheat by the natural selection of gene complexes conferring competitive ability in mixed communities.

Table 14.4
The harvest index of tillered, de-tillered and uniculm lines of wheat and barley.

Field crop	Location and years	Tillered cultivar	Tillered cv. restricted to two culms	Uniculm line
Wheat (Atsmon and Jacobs, 1979)	Tel Aviv, Israel Winter season	0.40		0.55
Wheat (Islam and Sedgley 1980)	Merredin, W. Aust.			
	1978	0.35	0.39	
	1979	0.24	0.30	
Barley[†] (Donald, 1979)	Roseworthy, S. Aust.			
	1975	0.45		0.54
	1976	0.41		0.51
	1977	0.40		0.45

† Values for a tillered cultivar (Clipper) at its optimum seed rate in each season (75 kg/ha) and for a uniculm barley (WID-101) sown each year at 112 kg/ha.

The attempts reported in this paper to develop the 1968 ideotype have not yet given commercial cultivars. The barley lines, for example, lack grain quality and adequate straw strength for the heavy ears. Nevertheless after ten years development towards the ideotype, the first main components of the design, the dwarf uniculm habit with high harvest index, seem to offer prospects of increases in wheat and barley yields of 15% or perhaps a good deal more. The second principal feature, reduced foliage and erect leaves, is still to be combined effectively with the non-tillering habit.

Concluding comment

It is proposed that communal wheat plants with a high harvest index may offer a route to substantially increased grain yields. The development of such plants calls for the reversal of certain evolutionary processes that have been operating continuously during the domestication and cultivation of wheat.

Within breeding programmes there is need to recognise that communal plants are unlikely to perform well if individual plant yield is a criterion of merit, or if planting arrangements favour competitive individuals. On the contrary, wherever feasible, the indices of performance should relate to the crop situation, in which there is intense competition among like individuals. Wheat breeding should be regarded not only as 'plant breeding' but also as purposeful 'crop breeding'.

References
Allard, R.W. and Jain, S.K. (1962) Population studies in predominantly self-pollinated species. II. Analysis of quantitative genetic changes in a bulk-hybrid population of barley. *Evolution* **16**, 90–191.

Allard, R.W., Jain, S.K. and Workman, P.L. (1968) The genetics of inbreeding populations. *Adv. Genet.* **14**, 55–131.

Athwal, D.S. (971) Semidwarf rice and wheat in global food needs. *Q. Rev. Biol.* **46**, 1–34.

Atsmon, D. and Jacobs, E. (1977) A newly bred 'Gigas' form of bread wheat (*Triticum aestivum* L.) : Morphological features and thermo-photoperiodic responses. *Crop Sci.* **17**, 31–35.

Bremner, P.M. (1969) Effects of time and rate of nitrogen application on tillering, 'sharp eyespot' (*Rhizoctonia solani*) and yield in winter wheat. *J. Agric. Sci.* **72**, 273–280.

Christian, C.S. and Gray, S.G. (1941) Interplant competition in mixed wheat populations and its relation to single plant selection. *C.S.I.R. Journal* **14**, 59–68.

CIMMYT Review (1979) The International Centre for Wheat and Maize Improvement, Mexico City.

Darwin, C. (1868) 'The Variation of Animals and Plants Under Domestication.' John Murray, London.

deBeer, G. (1958) Foreword in 'Evolution by Natural Selection'. Cambridge University Press.

Donald, C.M. (1962) In search of yield. *J. Aust. Inst. Agric. Sci.* **28**, 171–178.

Donald, C.M. (1968) The breeding of crop ideotypes. *Euphytica* **17**, 385–403.

Donald, C.M. (1979) A barley breeding programme based on an ideotype. *J. Agric. Sci.* **93**, 261–269.

Donald, C.M. and Hamblin, J. (1976) The biological yield and harvest index of cereals as agronomic and plant breeding criteria. *Adv. Agron.* **28**, 361–405.

Evans, L.T. and Dunstone, R.L. (1970) Some physiological aspects of evolution in wheat. *Aust. J. Biol. Sci.* **23**, 725–741.

Evans, L.T. and Wardlaw, I.F. (1976) Aspects of the comparative physiology of grain yield in cereals. *Adv. Agron.* **28**, 301–359.

Fischer, R.A. (1978) Are your results confounded by intergenotypic competition? Proc. 5th Int. Wheat Genetics Symp., New Delhi, 1978. pp. 767–777.

Grafius, J.E., Thomas, R.L. and Barnard, J. (1976) Effect of parental component complementation on yield and components of yield in barley. *Crop Sci.* **16**, 673–677.

Hamblin, J. and Donald, C.M. (1974) The relationships between plant form, competitive ability and grain yield in a barley cross. *Euphytica* **23**, 535–542.

Harlan, J.R. (1975) 'Crops and Man.' American Society of Agronomy, Madison, Wisconsin.

Harlan, J.R. (1976) In 'Evolution of Crop Plants'. (Ed. N.W. Simmonds.) Longman, London. pp. 93–98.

Harlan, H.V. and Martini, M.L. (1938) The effect of natural selection in a mixture of barley varieties. *J. Agric. Res.* **57**, 189–199.

Harlan, J.R., deWet, J.M.J. and Price, E.G. (1973) Comparative physiology of cereals. *Evolution* **27**, 311–325.

Islam, T.M.T. and Sedgley, R.H. (1980) in press.

Jensen, N.F. and Federer, W.T. (1964) Adjacent row competition in wheat. *Crop Sci.* **4**, 641–645.

Jensen, N.F. and Federer, W.T. (1965) Competing ability in wheat. *Crop Sci.* **5**, 449–452.

Johnston, R.P. (1972) Single plant yield as a selection criterion in barley. Thesis, University of Adelaide.

Jones, H.G. and Kirby, E.J.M. (1977) Effects of manipulation of number of tillers and water supply on grain yield in barley. *J. Agric. Sci.* **88**, 391–397.

Khalifa, M.A. and Qualset, C.O. (1975) Intergenotypic competition between tall and dwarf wheats. II. In hybrid bulks. *Crop Sci.* **15**, 640–644.

Kirby, E.J.M. and Jones, H.G. (1977) The relations between the main shoot and tillers in barley plants. *J. Agric. Sci.* **88**, 381–389.

Kranz, A.R. (1966) Stoffproduktion und Assimilationsleistung in der Evolution der Kulturpflanzen. II. Versuchsergebnisse und zusammenfassende Diskussion. *Biol. Zentralbl.* **85**, 681–734.

Law, C.N., Snape, J.W. and Warland, A.J. (1978) The genetical relationship between height and yield in wheat. *Heredity* **40**, 133–151.

Lee, J.A. (1960) A study of plant competition in relation to development. *Evolution* **14**, 18–28.

MacKey, J. (1966) The wheat plant as a model in adaptation to high productivity under different environments. Proc. 5th Yugoslav. Symp. on Wheat, Novi Sad. pp. 37–48.

MacKey, J. (1978) Wheat domestication as a shoot : root interrelation process. Proc. 5th Int. Wheat Genetics Symp., New Delhi, 1978.

Mayr, E. (1949) Speciation and systematics. In 'Genetics, Paleontology and Evolution'. (Eds. Jepsen, E.L., Mayr, E. and Simpson, G.G.) Princeton University Press, Princeton. pp. 281–298.

Nicholson, A.J. (1962) Population dynamics and natural selection. In 'The Evolution of Living Organisms' (Ed. G.W. Leeper.) Melbourne University Press.

Passioura, J.B. (1972) The effect of root geometry on the yield of wheat growing on stored water. *Aust. J. Agric. Res.* **23**, 745–752.

Percival, J. (1921) 'The Wheat Plant.' Duckworth, London.

Puckridge, D.W. and Donald, C.M. (1967) Competition among wheat plants sown at a wide range of densities. *Aust. J. Agric. Res.* **18**, 193–211.

Schultz, J.E. (1971) Soil water changes under fallow-crop treatments in relation to soil type, rainfall and yield of wheat. *Aust. J. Exp. Agric. Anim. Husb.* **11**, 236–242.

Soetono and Donald, C.M. (1980) Emergence, growth and dominance in drilled and square-planted barley crops. *Aust. J. Agric. Res.* **31**, 455–470.

Stebbins, L.G. (1974) 'Flowering Plants. Evolution Above the Species Level.' Harvard University Press, Cambridge Mass.

Suneson, C.A. (1949) Survival of four barley varieties in a mixture. *Agron. J.* **41**, 459–461.

Tanner, J.W., Gardener, C.J., Stoskopf, N.C. and Reinbergs, E. (1966) Some observations on upright-leaf-type small grains. *Can. J. Plant Sci.* **46**, 690.

Vogel, O.A., Allen, R.E. and Peterson, C.J. (1963) Plant and performance characteristics of semidwarf winter wheats producing most efficiently in Eastern Washington. *Agron. J.* **55**, 397–398.

Walter, C.J. and Barley, K.P. (1974) The depletion of soil water by wheat at low, intermediate and high rates of seeding. Proc. 10th Int. Cong. Soil Sci., Moscow, **1**, 150–158.

Wiebe, G.A., Petr, F.C. and Stevens, H. (1963) Interplant competition between barley genotypes. *N.A.S.N.R.C. Publ.* **982**, 546–557.

15 DEVELOPMENTS IN WHEAT AGRONOMY

R.A. Fischer

During domestication and subsequently, genetic change has been intimately associated with change in crop husbandry. Spectacular recent gains in productivity, for example, depend upon a strong positive interaction between improved short genotypes and higher levels of most agronomic inputs. It is therefore fitting that some explicit attention is given to the Cinderella of the applied sciences, crop agronomy and its interaction with wheat breeding.

Current status of wheat agronomy

Norman (1980) defined agronomy as "the science of manipulating the crop/environment complex with the dual aims of improving agricultural productivity and gaining a deeper understanding of the processes involved". To begin however it is useful to take a broader view of wheat agronomy by examining the general features of wheat production in the world today. The area sown to wheat annually now exceeds 230 Mha, some 60% in developed countries and 40% in, so called, lesser developed countries (LDCs). While acreage in the former has remained fairly stable over the last two decades, it has grown steadily at 1.6% per year in LDCs. The world wheat yield averaged 1.70 t ha^{-1} in the five year period 1974-1978, the corresponding average for developed countries being 1.97 t ha^{-1} and for LDCs 1.33 t ha^{-1} (Figure 15.1). Over the 20 years, yield has grown at 2.8% per year in both groups of countries. Taken along with area increases in the developing countries, world wheat production has grown at 3.2% per year, to reach an estimated 450 Mt in 1978-79, almost three times production 30 years ago. This is impressive growth for an enterprise which appears to depend so much on the vicissitudes of nature, and it is as much a credit to the wheat agronomist as it is to the wheat breeder.

The agronomic practices for wheat depend on the agroecological context in which the crop is grown. Six major agroecological situations for wheat production are evident when producing areas of similar characteristics are grouped (Table 15.1). From the breeding point of view, the first situation involves daylength-sensitive spring wheats, the next two involve strongly vernalization-requiring winter wheats, while the last three use wheats with little or no response to vernalization and little response to daylength. For the agronomist a major contrast emerges between semi-arid rainfed situations where wheat

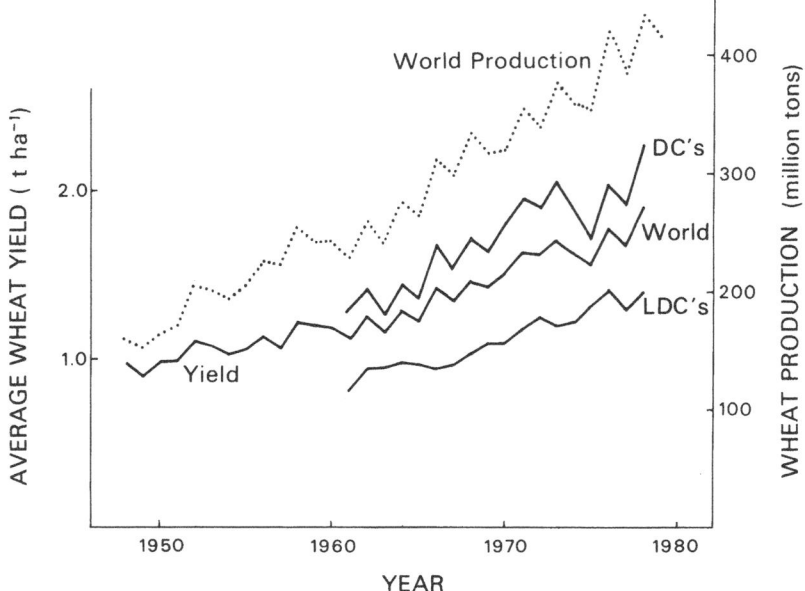

Fig. 15.1. Variation in average wheat yield for the world, developed countries (DCs) and lesser developed countries (LDCs). Source FAO.

commonly follows a long (1 year) fallow, and humid or irrigated areas where continuous arable cropping occurs. Grazing animals play a major part in the wheat farming system only in the intermediate latitude situation 4 of Table 15.1. Use of grazed leguminous pastures in most of the Australian wheatbelt for maintenance of stable soil nitrogen contrasts with the reliance on fertilizer nitrogen in most other developed countries.

As far as LDCs are concerned, the low latitude irrigated environments (the 5th agroecological context of Table 15.1) are of major importance. This is also the cradle of the 'green revolution' in wheat production and a brief look at a representative country in this situation is warranted. Since 1966, wheat yields in Pakistan have risen 75%. This increase in yield is undoubtedly closely related to the adoption of semidwarf or high yielding cultivars, estimated to occupy 75% of the area in 1977, to the steady increase in use of nitrogenous fertilizer and, finally, to associated improvement in other aspects of crop agronomy (water management, weed control, etc.).

Pressures for agronomic change

Changes in wheat agronomy are stimulated by changes in the availability and cost of inputs required for cropping, and by new technology. The most basic

Table 15.1

Characteristics of major agroecological situations in which wheat is grown in the world[a]

Latitude	Climatic features	Sowing period	Water supply to crop	Rotation followed	Countries or regions in which situation dominates
1. >45	Very cold winter, semi-arid	Spring	Rainfed	Wheat-fallow	Canada, central USSR
2. 35-50	Cold winter, semi-arid	Autumn	Rainfed	Wheat-fallow	USA, western USSR, Turkey, Iran
3. 40-60	Cold winter, humid	Autumn	Rainfed	Wheat - other annual crops	Western Europe, Eastern Europe
4. 25-40	Mild winter, semi-arid	Early winter	Rainfed	Wheat-fallow- grazed pasture	Mediterranean, Australia, Argentina
5. 15-35	Mild winter, semi-arid	Early winter	Irrigated	Wheat-other annual crops	India, Pakistan, Mexico, Egypt plus many other LDCs
6. 0-15	Cool summer, humid	Early summer	Rainfed	Wheat-other annual crops	Kenya, Ethiopia, Colombia, Ecuador

[a] China is excluded; it appears to contain elements of situations 1,3,4 and 5.

input of cropping is land. A recent report of the International Food Policy Research Institute (IFPRI, Oram *et al.*, 1979) concluded that wheat-producing LDCs, with the exception of Sudan, had little scope for further expansion of cultivated land. At the same time land is being lost through urbanization, erosion and salinization. All these factors will press for higher productivity from existing wheat lands and, where possible, extension of irrigation to rainfed regions, and intensification of cropping frequency.

The relative prices of the variable inputs into wheat production (i.e. labour, fuel, fertilizer, etc.) are also changing, particularly as a result of rising energy prices but also because of events such as the world fertilizer shortage in the mid 1970's. Over the last decade, statistics suggest that seed and fuel prices have inflated fastest, with fertilizer and machinery prices not far behind (FAO). Interestingly, plant protectants (fungicides, and presumably herbicides) have increased least. Australian data confirm that herbicides have shown the smallest price rises, and labour the largest (BAE, 1980).

In order to assess the impact of input price changes on agronomic practice, the relative amounts of inputs must be considered. These vary greatly with region but, by way of illustration, the relative expenditure on variable inputs in Oregon indicates that, despite substantial mechanization, labour is the biggest variable input and, surprisingly, fertilizers and chemicals each cost no more than seed (Cook *et al.*, 1979). However for the Oregon farm as a whole, the fixed costs per hectare sown (i.e. interest and depreciation on machinery and land) actually exceed the total cost of all variable inputs.

In future, liquid fuel prices will presumably continue to increase substantially and inputs based on them may be at a disadvantage. Nitrogenous fertilizer, for example, requires approximately 2 *l* diesel equivalent per kg N. The lower energy requiring inputs are herbicides at approximately 1 (2,4-D) to 5 (paraquat, glyphosphate) *l* diesel equivalent for the chemical needed to treat 1 hectare (Green and McCulloch, 1976), phosphorus fertilizer (0.7 *l* diesel equivalent per kg P), seed, and labour. An example of the alternatives which could be facing wheat farmers is given by Ellington (1979) who reported a comparison of conventional cultivation and direct drilling in northern Victoria. The former strategy required 16 litres of diesel and 2 hours of labour per hectare sown, the latter 4.5 litres of diesel, 0.7 hours of labour, and one application of paraquat/diquat per hectare. At 1979 prices, the total costs of these inputs were equal for each strategy.

Input cost inflation, seemingly devastating when viewed alone, should be related to product price inflation. Obviously in the long term there is a close connection between the two, as is further illustrated for a key ratio in world wheat production, that of the cost of fertilizer nitrogen relative to the price of wheat. Despite inflation, this ratio has changed little over the last decade, and it is unlikely that the relative cost of nitrogen (presently 3 kg wheat per kg N in USA) will rise sufficiently to preclude its use on wheat for, on a global basis, cost rises will be passed on to the wheat consumer. It is also worthwhile noting that the relative cost of diesel fuel (presently 3 kg wheat per litre of diesel fuel in Australia) is still low in view of fuel productivity in wheat cropping (about 30 kg wheat l^{-1}).

Concern for environmental pollution and for consumer protection from agrochemicals is another force influencing the availability and cost of inputs. Certain practices are or could be enforced by law (e.g. stubble retention for erosion control). Some pesticides and herbicides might be banned, and the cost of new ones reflects this heightened concern. But such legislation will make for better chemicals, not for chemical-free agronomy.

Before examining new technology as a source of agronomic change, I would like to emphasize the great improvements in wheat husbandry at the farm level which are to be achieved through the more widespread adoption of existing technology. The research station-farm gap still exists, especially, but not only, in LDCs where closing the gap offers the greatest hope for productivity gains. New high yielding varieties play a role here for, although their original development probably required the stimulus of agronomic change, particularly the use of cheap N fertilizer, their spread to new regions has in turn stimulated the simultaneous adoption of an 'improved agronomic package', because of its readily demonstrable economic advantages.

Since I began by distinguishing developed and lesser developed countries, it might seem that the consideration of new technology should be similarly divided. As Evans (1976) pointed out, it has become fashionable to praise the low input, apparently self-sufficient and non-polluting agriculture of undeveloped communities, and at the same time condemn the application of modern high input western agricultural techniques in such situations. The energy crisis has reinforced this popular view, but I would suggest that the anticipated food crisis of the next 20 years is of much more immediate concern to the LDCs themselves. Also I believe that except for certain considerations of the relative availabilities of labour and capital, the optimal agronomy for wheat in LDCs will not differ from that in developed countries except in so far as the physical environments such as latitude and soils differ. The terms low and high input should refer to the productive potential of the particular environment, meaning climatic environment, which is largely impossible to modify, and aspects of the soil environment, which are too expensive to modify (e.g. lack of profile depth or extreme acidity). LDCs, least of all, can afford low input agronomy in potentially productive environments. Whether they can afford, or will be given access to, the capital needed to lift crop production is another question; the IFPRI report estimates that about $US500 needs to be invested in irrigation, mechanization, fertilizer plants etc. for each one ton increase in sustainable annual food production (Oram *et al.*, 1979).

Crop Nutrition and Soil Toxicities
Fertilizers

The importance of fertilizers for sustained production is illustrated by wheat yields in the continuous cropping experiments laid down by Lawes and Gilbert in Broadbalk field at Rothamsted in 1843. Unfertilized winter wheat today yields 1.7 t ha^{-1} compared to 5.5 t ha^{-1} for that receiving annual dressings of combined chemical fertilizer, including 144 kg ha^{-1} N (Johnston and Mattingly 1976). Calculations in the IFPRI report assume that increase in grain production by LDCs will require 1 kg of total nutrients in the form of fertilizer for each 8 (rainfed) to 10 (irrigated) kg of grain. For nitrogen alone, a ratio between 1:20 and 1:30 is required. This inescapable need for fertilizer at a time of sharply rising energy and fertilizer costs has forced attention to the efficiency of fertilizer utilization.

In the case of N, new slow-release forms or nitrification inhibitors are useful in situations where losses may be great between application and crop uptake e.g. the inhibitor nitrapyrin doubled the yield response to autumn applied N in winter wheat in Indiana (Huber *et al.*, 1980). Split application of N (e.g. part at sowing and the rest at jointing) offers another way of reducing such losses, and also of adjusting N supply to likely water availability in rainfed situations. There will be renewed interest in lower grade forms of P, such as rock phosphate, but with its slow rate of supply of available P, rock phosphate is likely to remain unsuited for meeting the P demands for rapid growth of wheat. In order to reduce transportation costs and for ease of bulk handling we see more compound forms of fertilizer, higher analysis fertilizers, and in U.S.A. the appearance of concentrated liquid fertilizers, mixed to the customer's requirements at the railhead.

Nutrient-efficient cultivars

For the reasons given above, there has been considerable recent interest in genetic variation in efficiency of nutrient use (E), the yield of grain per unit of supply of limiting nutrient. E can be usefully considered as the product of nutrient uptake per unit of supply (uptake efficiency, E_u) and grain yield per unit of uptake (utilization efficiency, E_z):

$$E = E_u \times E_z \qquad (1)$$

E_u is usually less than 1 because of nutrient losses from the soil and fixation of fertilizer nutrients in the soil. If losses are substantial, as in the case of nitrogen leaching and denitrification in wet situations, genotypic strategies to reduce losses could be devised. For example rapid early root growth could secure more of the labile soil N before loss-inducing winter rains occurs. Also, deeper rooting may permit access to deeply leached nitrogen. Where nutrients are fixed chemically by the soil, the ability to exploit less available forms of nutrients would increase uptake efficiency. For example extensive rooting, mycorrhizal associations and/or root chemical properties may improve exploitation of fixed phosphorus.

An example of a phosphorus-efficient variety is known in rice, where more extensive rooting appears to be the cause of greater efficiency, conferring on the variety a specific advantage when P is limiting (Koyama and Chammek, 1971). However, examples of such differences between crop cultivars are as yet rare (Salinas and Sanchez 1976), and even populations of *Anthoxanthum odoratum* L. grown for 112 years with and without applied P in the Park Grass Experiment at Rothamsted showed no specific adaptation to low P in sand culture, for the high P population produced most dry matter at all P levels (Davies and Snaydon, 1974).

Theoretically better exploitation of fixed phosphorus can result in reduced P applications only until residual P in the soil is built up sufficiently that only maintenance dressings are required. I estimate that for the average Australian wheat farm we are currently removing in produce (largely wheat) about 40% of the P supplied as fertilizer each year. When we will reach a maintenance situation is unclear. Where the soil is a long way from this position and where its total P content is high but available P low (e.g. volcanic ash soils or andesols in Chile which require more than 50 kg/ha of P per year for reasonable wheat yields), cultivars better able to exploit fixed P would be most desirable.

In the case of minor and trace elements, total soil content is usually large relative to crop removal. Genetic increases in uptake efficiency would be useful in situations of incipient deficiency. Rice cultivars less sensitive to Zn and Fe deficiency are known. Recently it was shown that some triticales, like their rye parents, are less sensitive to Cu deficiency this being due to one or several rye chromosomes conferring greater Cu uptake (Graham and Pearce, 1979).

Efficiency of utilization of nutrients (E_z) can be considered as follows:

$$E_z = HI_n / [G]_n \qquad (2)$$

where HI_n is the harvest index of the particular nutrient (i.e. the percentage of total uptake in the grain at maturity) and $[G]_n$, the percentage concentration of the nutrient in the grain. Typical values for nitrogen in wheat are $HI_n = 75\%$ and $[G]_n = 2.5\%$, giving $E_z = 30$ kg/kg. However, breeding for higher E_z must face several problems:

(i) Genotype differences are likely to be greatest in those environments which maximize E_z. Inevitably, with major nutrients these are environments in which the particular element is severely deficient and yield is likely to be very low (the same problem also applies to breeding for high E_u). In other words efficiency mechanisms are likely to be of importance at yield levels too low to be economic.

(ii) Increased E_z would result from increased nutrient harvest index (HI_n), but for N and P (and probably the other elements) highest HI_n values are already over 75%. Further increases may be counter- productive in terms of yield if, as is probable in the case of nitrogen, it involves faster removal from the leaves of an element essential to the photosynthetic system.

(iii) Increased E_z through reduced nutrient concentration in the grain ($[G]_n$) would also be problematical in the case of nitrogen, since it implies reduced % protein in the grain. However, reduced grain phosphorus may be tolerable since much grain P is in the storage compounds, phytic acid and derivatives, which are largely removed in the bran fraction with milling and are of negative nutritive value for monogastric animals (Hulse and Laing, 1974). Australian wheats presently contain about 0.3% P in the grain, a low value by world standards (0.4% being about average), presumably because of the poor soil P status in Australia.

I conclude that breeding specifically for nutrient use efficiency is going to be difficult, with little likelihood of success, although there are some possibilities in the case of efficiency of P and microelement use. This conclusion should not obscure the fact that the efficiency, as I have defined it, of modern cultivars is higher than that of the old cultivars. This is the indirect result of increased yield potential, as illustrated for nitrogen in Figure 15.2. In practice this gain in efficiency is not fully exploited as such; rather, because of economic considerations, fertilizer rates have increased for the high yield potential varieties, such that approximately equal efficiencies of N use are reached, but at higher yield levels. The increase in N efficiency at higher levels of nitrogen input with the shorter, higher yield potential cultivars has resulted less from increased nutrient uptake (E_u) than from increased efficiency of utilization (E_z), due both to increasing nutrient harvest index and decreasing grain N content (Table 15.2).

Other sources of nitrogen

While much of the world's food grain production will continue to be driven by fertilizer nitrogen, other factors encourage reliance on non-fertilizer nitrogen sources in certain situations, most notably in Australian agriculture. Traditionally the grazed leguminous pasture has provided sufficient soil nitrogen for the Australian wheat crop. Attempts to introduce this system into North Africa

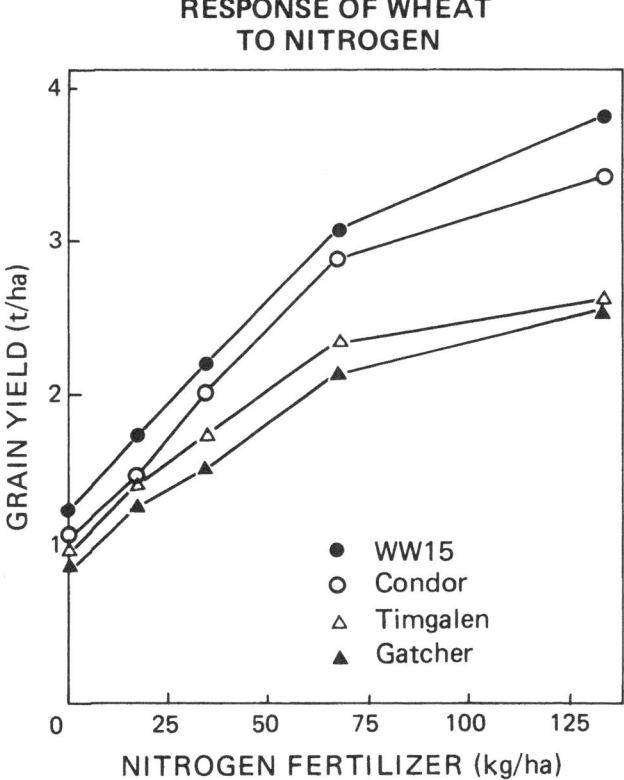

RESPONSE OF WHEAT
TO NITROGEN

Source: Syme *et al.* (1976)

Fig. 15.2. Grain yield response of two semidwarf (WW15, Condor) and two non-dwarf (Timgalen, Gatcher) wheat cultivars to fertilizer nitrogen; rainfed, southern Queensland (Syme *et al.*, 1976).

have demonstrated its technical feasibility (Breth, 1975), but socioeconomic barriers remain. More recently in Australia, grain legumes such as lupins and peas are receiving attention as sources of N within the wheat rotation. Vigorous lupin crops appear to fix enough nitrogen that, even after their grain is removed, there is a net gain in soil nitrogen. Peas, vetches and clover-based green manures are used in humid areas like Europe, while other grain legumes such as lentils, chick peas and *Vicia faba* are rotated with wheat in traditional cropping systems of the Middle East and the Indian sub-continent. Apart from nitrogen fixed, such crops also offer advantages in weed and disease control, but more research

Table 15.2

Grain yield, total nitrogen uptake, nitrogen harvest index, grain nitrogen concentration and efficiency of nitrogen utilization (kg grain/kg N uptake) for spring wheat cultivars grown under irrigation and high fertility in northwest Mexico, 1970-71 (R.A. Fischer, unpublished).

Cultivar group (n)	Grain yield (t ha^{-1})	Total N uptake (kg ha^{-1})	N harvest index (%)	Grain N (%)	Efficiency of N utilization (kg ha^{-1})
Tall (5)	5.76	200	67.6	2.72	28.3
Low yielding semi dwarf (9)	6.33	206	70.7	2.62	30.8
High yielding semi dwarf (10)	7.10	211	72.4	2.45	33.8
S.E.	0.13	4	1.0	0.05	0.6

is needed before they could replace the grazed leguminous pasture in sustainable cropping systems which do not rely on N fertilizer. Pressure on the pasture phase in Australia is more likely to come from the recent trend for greater use of fertilizer nitrogen to increase the proportion of land cropped to cereals.

The implications of these possible changes for breeding are minimal. Some have suggested that wheat and legumes should be intercropped for maximum efficiency, and cultivars suited for intercropping developed. However, given that wheat usually already fills the available growing season in semi-arid areas, and therefore that critical phases for wheat and any interplanted legume are likely to be coincident, yield advantages in such mixtures appear biologically unlikely, not to mention practical difficulties with weed control and mechanization.

There has been recent interest in non-leguminous sources of biologically-fixed nitrogen in cropping systems. In the case of wheat the best possibility seems to be the use of the left-over straw to stimulate N fixation by free-living soil microbes. Significant fixation has been measured when straw-soil mixtures are moistened, some wheat soils being more active than others. Decomposition of cellulose and lignin of straw must precede the flush of fixation; more readily available carbohydrates in straw (normally < 5% of the dry weight) should facilitate the process. However much more work is needed before this source of nitrogen can be considered significant.

Soil toxicities

Some soils are naturally very acid and others of low cation exchange capacity in humid areas are prone to acidification under nitrogen-fertilized cropping or, or as is being found in Australia, phosphorus- fertilized leguminous pastures. Low soil acidity can lead to levels of aluminium and manganese toxic to wheat and many other species. While only liming can arrest or reverse pH decline, useful genetic differences in sensitivity to both aluminium and manganese exist between wheat genotypes. For example, in the Oxisols of southern Brazil, introduced wheat varieties without resistance die while adapted local cultivars yield 1–2 t ha^{-1}, and less lime is required to maximize the yield of the latter cultivars. The cost of liming and the rapid expansion of areas affected by the problem mean that breeders will have to give more attention to genetic resistance to low pH — induced toxicities ; fortunately simple screening for resistance is feasible.

In other more arid regions increasing salinity of wheat soils is a problem and this is often associated with a rising water table. Satisfactory genetic differences in salinity resistance have yet to be demonstrated. Also, inexorable accumulation of salt at the soil surface ultimately demands agronomic or engineering solutions which reduce the net recharge of water tables and/or permit washing of the soil. Genetic resistance, if it exists, may be useful where a saline equilibrium situation prevails.

Land preparation and tillage
Fallowing

Although the statement by Jethro Tull that 'tillage is manure' may no longer be considered relevant, that of H.W. Campbell asserting that 'tillage is moisture' still holds much force. Campbell was the advocate of 'dry farming' in the Great Plains of North America at the turn of the century. As a consequence, where annual rainfall is less than about 550 mm in the Great Plains and 450 mm in the Pacific North West, wheat is sown after a 14–17 month (winter wheat) or 19–21 month (spring wheat) fallow period of weed control. However the original clean cultivated fallow is being replaced by stubble mulch tillage methods which retain as much crop residue as possible on the surface, thereby improving soil erosion control and, especially in areas of summer precipitation, improving soil water storage (Fenster *et al.*, 1977; Van Doren and Allmaras, 1978). In high latitude wheat regions (situations 1 and 2 of Table 15.1) when average rainfall is less than about 400–500 mm, winter wheat after fallow on average yields more than twice that of continuous wheat; also annual costs of the wheat fallow system are lower (Oregon State University, 1979). A very important feature of this system is the maintenance of soil moisture close to the soil surface and/or the use of deep furrow drills, so that early sowing of winter wheat onto residual moisture, well before the onset of winter cold, is ensured.

Given the grain yields obtained, the energy-efficient tillage techniques now developed (e.g. rod weeder, sweep cultivators), and the use of crop residue to control erosion, the long or summer fallow system is considered relatively stable and is unlikely to be discarded in favour of continuous cropping. It is also probably the optimal system for semi-arid parts of western Asia such as the Anatolian plateau. The wheat fallow system is, however, inefficient in that approximately three quarters of the precipitation of the fallow period is lost through soil evaporation. It remains a challenge to the agronomist to reduce this loss further or to take advantage of it.

Crop residue retention tends to reduce soil temperature, and cause phyto-toxicity in some situations. Also certain wheat diseases may be encouraged by carryover on residue or by the microclimate provided (Cook et al., 1978). Carryover is likely to favour above-ground pathogens like Septoria spp. and yellow leaf spot, Pyrenophora tritici-repentis; microclimatic changes may favour root and crown pathogens such as take-all (Gaeumannomyces graminis var. tritici), and eyespot lodging (Cercosporella herpotrichoides). Yellow leaf spot may become a serious hazard of stubble mulching in north eastern Australia (Rees and Platz, 1979). Residue retention is undoubtedly the way of the future, and breeders must give more attention to the incorporation of genetic resistance against the pathogens likely to be encouraged.

Some interest is now being taken in chemical fallowing and a hybrid system involving early chemical application and late mechanical tillage, aptly termed 'eco-fallow', has been developed in western Nebraska. The chemical fallow, in terms of moisture conservation, is at least equal (eastern Oregon, F.E. Bolton, personal communication), if not better than (high plains of Texas, Unger, 1978; Canadian Prairie, Downing, 1977), residue-retention tillage. It may pose a problem where early sowing on deep residual moisture is required for maximum yields, but new machines for deep drilling into the undisturbed seed bed are being developed (e.g. the strip tiller of Bolton and Booster, 1979). It is probable that cultivars will need genetic resistance to the residues of the soil-persistent herbicides best suited for chemical fallowing.

When one turns to Australian agriculture, or to lower latitude wheat regions in general, it appears that the long fallow is not adhered to so rigidly. In fact fallowing has been on a steady decline in Australia, where only 11% of the wheat area was long-fallowed in 1978–1979. One reason appears to be that fallowing conserves less moisture here than it does overseas, partly because of less suitable soils (less retentive subsoils, hard-setting shallow top soils). French (1978a) measured an average storage of 28 mm in South Australia. Also, the warmer winters of intermediate latitudes mean more crop growth during the cool wet period of commonly greatest rainfall when transpiration efficiency is high (Fischer, 1979). Growing season rainfall of 400 mm gave yields of 2200 kg ha^{-1} in the South Australian study (French, 1978b), but less than 1000 kg ha^{-1} with continuous winter wheat in the Great Plains (Oregon State University, 1979). Other reasons which have facilitated the demise of long fallow in Australia include improved soil nitrogen through the use of leguminous pastures, the availability of animals on the typical mixed wheat farm to take advantage of the extra pasture growth obtained by not fallowing, and the advent of herbicides for better weed control. Wheat cultivars resistant to root rots would remove another reason often given to justify the long fallow.

Direct Drilling

In southern Australia the trend away from fallow is, in fact, leading to considerable interest in zero-till or direct drilling. Direct drilling refers to sowing into previously uncultivated soil, usually following spraying with a contact broad-spectrum herbicide to kill growing weeds; zero-till implies in addition minimal soil disturbance with the sowing operation. Some 100 000 ha were direct drilled in Western Australia last year, and 36 000 ha in southern New South Wales this season (1980), more than double last year's area. Also the amount of direct drill or zero-till wheat has been steadily expanding in humid regions overseas, such as the United Kingdom (Cannell *et al.*, 1978), south-eastern U.S.A. (Phillips *et al.*, 1980), and southern Brazil (Barker and Wunsche, 1977).

The general advantages of drilling directly rather than after prior ploughing and cultivation include greatly reduced erosion hazard, substantial reduction in liquid fuel consumption and reduced total energy input, reduced labour costs and increased timeliness and flexibility of cropping (Cannell and Ellis, 1979; Phillips *et al.*, 1980). In humid areas direct drilling may so speed the turn-about between crops that double cropping becomes a possibility where it wasn't beforehand (e.g. wheat/soybeans in southeast U.S.A. and southern Brazil). On mixed farms direct drilling increases the availability of pasture in the autumn when there is often a shortage of fodder.

Direct drilling sometimes gives lower yields than those obtained with a short cultivated fallow, an effect which seems to be associated with reduced early plant vigour. An example from local studies is shown in Figure 15.3; reduced early growth and final yield at Murrumbateman contrasts with equal growth and yield at Wagga Wagga. In the U.K. this problem is worse on certain soils but seems to disappear on others with continued direct drilling (Cannell and Ellis,

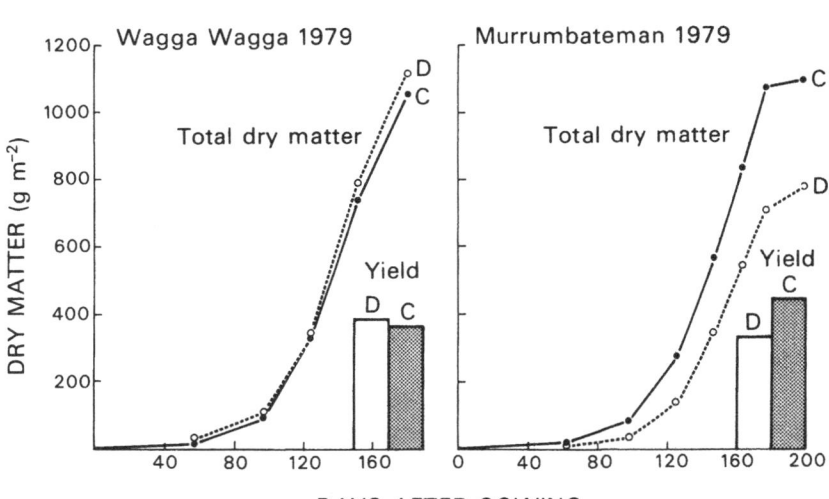

Fig. 15.3. The growth and yield of wheat (cv. Egret) either direct drilled (D) or sown after a conventional short cultivated fallow (C) at two sites following pasture in southern N.S.W. in 1979 (R.A. Fischer and colleagues, unpublished).

Table 15.3
Effect of plant residue mulch on yield of rainfed wheat in India

Location and treatment	Year	Growing season rainfall (mm)	Grain (t ha^{-1})		
			Control	Mulched	% increase
New Delhi[a]	73-74	28	0.85	1.07	26
	74-75	58	1.03	1.51	47
Ludhiana[b]	73-74	70	2.54	2.96	17
	74-75	120	1.53	1.99	30
	75-76	260	3.1	3.43	9

[a] Prihar *et al.* (1979)
[b] De and Giri (1978)

1979). The reduced vigour is probably related to a reduction in the volume of disturbed soil around the seedlings, root growth into the undisturbed region being too slow under certain adverse conditions (e.g. low soil moisture, hard soil type, low soil temperature, absence of old root channels).

The generally drier conditions in Australia and other intermediate latitude wheat regions, compared with humid regions elsewhere where direct drilling of wheat is proving successful, may pose special difficulties for direct drilling. In particular some presowing moisture conservation may be foregone by not cultivating, and sowing may therefore be delayed. Nevertheless direct drilling is giving yields equal to those with a short cultivated fallow in the Mediterranean conditions of south western Australia and in more humid parts of the south east (Reeves and Ellington, 1974; Rowell *et al.*, 1977). Also direct drilling permits the retention of plant residue on the soil surface both before and, if appropriate drills are designed, after sowing. In dry situations plant mulches can increase yields considerably through moisture conservation via reduction in direct soil evaporation, a major component of water loss in dryland crops (Table 15.3; Van Doren and Allmaras, 1978).

Although Australian farmers are using conventional tyne drills for direct drilling of wheat in favourable situations (soft soils, optimal moisture conditions), drills especially adapted to direct drilling are now available or are being tested. The strip tiller developed in Oregon and mentioned earlier is one example. The Siroseeder developed in CSIRO Division of Plant Industry is another. Both implements aim to prepare a slot for the seed, filled with loose soil and free of plant residue, while leaving the soil between the row undisturbed. The ability of machines to handle heavy plant residue will be important; the triple disc drill is a true zero-till implement and excels in cutting through residue but seems unsuited to dry conditions because little soil is returned to the slot. No doubt there will be considerable engineering advances in this area. A hint of things to come is seen in the strip tiller which incorporates injection of both water and nitrogenous fertilizer with the seed (500–1000 L water ha^{-1} can substantially improve germination in marginal conditions), and also precision application of a residual herbicide to the inter-row space.

I am of the firm opinion that reduced tillage, and particularly direct drilling, will be of major significance where wheat is grown without a long moisture-conserving fallow. We need much more research on fundamental aspects of the technique, but even with reduced yields it will be an attractive alternative to the farmer when its adoption means the possibility of double cropping (humid areas) or extra grazing animals (mixed farms of especially semi-arid areas) or less erosion (sloping lands). It will call for some accommodation by the breeders with possibly earlier cultivars and, in the longer term, cultivars suited to a largely undisturbed seed bed, perhaps covered with a plant residue mulch.

Weeds and herbicides

Weeds probably constitute the major agronomic problem for wheat growing around the world. In the past, mechanical cultivation prior to sowing has been the principal means of control, but the existence of dormant seeds of many weed species (e.g. *Avena fatua*) reduces its effectiveness and often causes delays in sowing, not to mention the cost of repeated cultivation. Preventive methods aimed at stopping weeds from propagating in the years prior to cropping (long fallow, hay cutting, etc.) are useful but of similar limited effectiveness.

Herbicide use on wheat began some 40 years ago and has steadily grown, with a marked increase in the number, specificity and crop stages for application of available herbicides. The price of herbicides has not increased as fast as that of other inputs and the energy cost of manufacturing the herbicide needed to treat one hectare is relatively low. Notwithstanding public concern about herbicides, I expect their use to continue to increase because they generally represent the cheapest means of controlling weeds. Also once the wheat crop has been planted, they are the only means of weed control. It is likely that the safety and effectiveness of herbicides will also continue to improve, with much scope for better means of applying herbicides (Swarbrick, 1980) e.g. controlled droplet application, residual herbicides not requiring soil incorporation, more versatile tank mixes, herbicide additives and antidotes.

Desirable selective herbicides often need to be applied at rates which may damage certain wheat cultivars. At the least, therefore, wheat breeders must screen new material against the probable herbicides to be encountered. Whether explicit breeding for herbicide resistance is warranted is another question. For example, resistance to widely used triazine herbicides has arisen spontaneously in a weedy race of *Brassica campestris* in Canada (Souza Machado *et al.*, 1978). Resistance in wheat to these or other broad-spectrum residual herbicides would be very useful, greatly facilitating the use of chemical fallowing. If genetic resistance cannot be located or generated with conventional methods, including screening in tissue culture, the construction and incorporation into wheat of a herbicide-resistance gene might provide one of the less complex, and hence remotely feasible, opportunities to prove the worth of recombinant DNA techniques.

From time to time there has been interest in differences in the ability of cultivars to compete with weeds. It is a general observation that the newer semidwarf wheats are not as competitive as older taller varieties. Reeves and Brooke (1977) compared 29 genotypes in competition with annual ryegrass (*Lolium rigidum*). The percentage yield reduction at 1500 ryegrass plants m^{-2} was significantly affected by cultivar, ranging from 23 to 48%, but the reduction was not related to genotype height, a result supported by agronomic research in Mexico (CIMMYT, 1976). Reeves and Brooke (1977) suggested that differences in early root development may have caused differences in competitiveness. Whatever the causes, such genotypic effects deserve more attention. High seeding densities and narrow rows also offer small gains in the competitiveness of the crop.

Competitive crops will never eliminate the need for herbicide where weed density is high, but could do so where density is low, or could usefully reduce weed seed production in the crop. Eradication of weeds is out of the question for most wheat farming systems and most weeds; the best strategy is to employ the minimal inputs possible to keep weed growth within the acceptable limits of detrimental effects on yield or grain cleanliness.

Other agronomic considerations

Space does not permit adequate attention to other important areas of the subject. These include the question of optimal sowing date, its interaction with maturity class of cultivar (Fischer, 1979), the key role of agronomic techniques which may ensure sowing at or near the optimal date (fallowing, deep furrow drills, press wheels, water injection), and the need for cultivars whose flowering date is less sensitive to inevitable variation in sowing date. Table 15.4 illustrates the importance of timely sowing.

A second area is the role of relatively simple simulation models of the wheat crop. When combined with historical weather records, they permit the testing of effects of altered agronomic practices on yield and yield stability. Modelling will not eliminate field experimentation, but it will make it more efficient. Simple models and minicomputers will also be used to aid decisions by farmers on agronomic practices (e.g. sowing date, seeding density, fertilizer rates in the face of actual or expected moisture supply, known soil fertility and expected prices).

A third development, seen particularly in the productive humid wheat areas of western Europe, is the increasing application of insecticides, fungicides and growth regulators to the wheat crop. While this development has been stimulated by the advent of varieties of very high yield potential and by generous grain prices, it seems to suggest that resistance breeding has not, or cannot, do the job e.g. for the control of *Septoria* spp., stripe rust and mildews.

Finally, crop rotation is an important part of wheat agronomy. The use to which the land is being put between wheat crops influences not only soil nitrogen and weed population but also levels of soil-borne diseases, soil moisture and soil structure. Where animals are present, grazing of the wheat crop itself, as well as the residue, may be important considerations. Also it is necessary to consider effects of the wheat crop and its residue on subsequent activities in the rotation (e.g. pasture establishment, erosion control). Wheat cropping is always

Table 15.4

The effect of sowing date on wheat yield

Location	System	Optimal sowing period	Decrease in relative yield per week delay after optimal period
Sholapur, India (18° N)[a]	Post-monsoon spring wheat	September	10%
New Delhi, India (28° N)[b]	Irrigated spring wheat	November	7%
Tamworth, Australia (31° S)[c]	Rainfed spring wheat	late June	6%
Wagga Wagga Australia (35° S)[d]	Rainfed spring wheat	early May	4%
Anatolian Plateau, Turkey (40° N)[e]	Winter wheat on long fallow	mid September to mid October	5%

[a] Krishnamoorty (1981)

[b] Bhardwaj *et al.* (1975)

[c] Doyle and Marcellos (1974)

[d] Kohn and Storrier (1970)

[e] Bolton (1974)

part of a farming system and the wheat agronomist, if not the breeder, must be aware of this. Only agronomic changes which maintain or improve the long term stability and productivity of the whole system are acceptable.

References
BAE (1980) 'Historical Trends in Australian Agricultural Production, Exports, Incomes and Prices.' Bureau of Agricultural Economics, Canberra.

Barker, M.R. and Wunsche, W.A. (1977) Plantio direto in Rio Grande do Sul, Brasil. *Outlook Agric.* **9**, 114—120.

Bhardwaj, R.B.L., Jain, N.K., Wright, W.C., Sharma, K.C., Gill, G.S. and Krantz, B.A. (1975) 'Agronomy of Dwarf Wheats.' Indian Council of Agricultural Research, New Delhi.

Bolton, F.E. (1974) Tillage and cultural practices for wheat under low rainfall conditions – soil and crop management research in the wheat research and training project, Turkey. Proc. 2nd Regional Wheat Workshop, Ankara, 1974

Bolton, F.E. and Booster, D.E. (1979) Strip-till planting for dryland cereal production. Paper prepared for presentation at 1979 Winter Meeting of American Society of Agricultural Engineers, New Orleans.

Breth, S.A. (1975) The return of medic. CIMMYT Today. International Maize and Wheat Improvement Centre, El Batan, Mexico.

Cannell, R.Q. and Ellis, F.B. (1979) Simplified cultivation: effects on soil conditions and crop yield. *Agric Res. Coun. Res. Rev.* 5, 55–59.

Cannell, R.Q., Davies, D.B., Mackney, D. and Pidgeon, J.D. (1978) The suitability of soils for sequential direct drilling of combine-harvested crops in Britain: a provisional classification. *Outlook Agric.* 9, 306–316.

Cook, R.J., Boosalis, M.G. and Doupnik, B. (1978) Influence of crop residues on plant diseases. In 'Crop Residue Management Systems'. (Ed. W.R. Oschwald.) American Society of Agronomy, Special Publication 31, pp. 147–163.

Cook, G.H., Holst, D.L. and Macnab, S. (1979) Estimated wheat production and marketing costs on a 2000–acre dryland farm, Oregon Columbia Plateau, 1979–1980. Special Report 528. Oregon State University Extension Service, Corvallis, Oregon.

Davies, M.S. and Snaydon, R.W. (1974) Physiological differences among populations of *Anthoxanthum odoratum* L. collected from the Park Grass Experiment, Rothamsted. *J. Appl. Ecol.* 11, 699–708.

De, R., and Giri, G. (1978) Effect of mulches and kaolin foliar spray on wheat yield in drylands. *Indian J. Agric. Sci.* 48, 334–336.

Downing, C.G.E. (1978) Mulch tillage practices and equipment for cereal crop production in Western Canada. Proc. Int. Conf. Energy Conservation in Crop Production. Massey University, Palmerston North. pp. 137–145.

Doyle, A.D. and Marcellos, H. (1974) Time of sowing and wheat yield in northern New South Wales. *Aust. J. Exp. Agric. Anim. Husb.* 14, 93–102.

Ellington, A. (1979) Energy requirements and operating costs of reduced vs. conventional tillage. Paper presented at Australian Agricultural Engineering Society Specialty Conference on Tillage, Narrabri, 1979.

Evans, L.T. (1976) The two agricultures: renewable or resourceful. *J. Aust. Inst. Agric. Sci.* 42, 222–231.

FAO Production Year Book. 1972, 1975, 1978. Food and Agriculture Organization, Rome.

Fenster, C.R., Owens, H.I. and Follett, R.H. (1977) 'Conservation Tillage for Wheat in the Great Plains.' U.S. Dept of Agriculture Extension Service.

Fischer, R.A. (1979) Growth and water limitation to dryland wheat yield in Australia: a physiological framework. *J. Aust. Inst. Agric. Sci.* 45, 83–94.

French, R.J. (1978a) The effect of fallowing on the yield of wheat I. The effect on soil water storage and nitrate supply. *Aust. J. Agric. Res.* 29, 653–668.

French, R.J. (1978b) The effect of fallowing on the yield of wheat II. The effect on grain yield. *Aust. J. Agric. Res.* 29, 669–684.

Graham, R.D. and Pearce, D.T. (1979) The sensitivity of hexaploid and octoploid triticales and their parent species to copper deficiency. *Aust J. Agric. Res.* 30, 791–799.

Green, M.B. and McCulloch, A. (1976) Energy considerations in the use of herbicides. *J. Sci. Food Agric.* 27, 95–100.

Huber, D.M., Warren, H.L., Nelson, D.W., Tsai, C.Y. and Shaner, G.E. (1980) Response of winter wheat to inhibiting nitrification of fall applied nitrogen. *Agron. J.* 72, 632–637.

Hulse, J.H. and Laing, E.M. (1974) 'Nutritive Value of Triticale Protein'. International Development Research Centre, Ottawa.

Johnston, A.E. and Mattingly, G.E.G. (1976) Experiments on the continuous growth of arable crops at Rothamsted and Woburn Experimental Stations: effects of treatments on crop yields and soil analyses and recent modifications in purpose and design. *Ann. Agron.* 27, 927–956.

Kohn, G.D. and Storrier, R.R. (1970) Time of sowing and wheat production in southern New South Wales. *Aust. J. Exp. Agric. Anim. Husb.* 10, 604–609.

Koyama, T. and Chammek, C. (1971) Soil-plant nutrition studies on tropical rice. I. Studies on the varietal differences in absorbing phosphorus from soil low in available phosphorus. *Soil Sci. Plant Nutr.* 17, 115–126.

Krishnamoorty, Ch. (1981) Low input cropping systems. Proc. Symp. Potential Productivity of Field Crops Under Different Environments. International Rice Research Institute, Los Banos (in press).

Norman, M.J.T. (1980). Your future depends on agronomy. *J. Aust. Inst. Agric. Sci.* 46, 105–111.

Oram, P., Zapata, J., Alibaruho, G. and Roy, S. (1979) Investment and input requirements for accelerating food production in low-income countries by 1990. Research Report 10. International Food Policy Research Institute, Washington D.C.

Oregon State University (1979) 'Dryland Agriculture in Winter Precipitation Regions of the World.' Dryland Agriculture Technical Committee, Oregon State University, Corvallis.

Phillips, R.E., Blevins, R.L., Thomas, G.W., Frye, W.W. and Phillips, S.H. (1980) No-tillage agriculture. *Science* 208, 1108–1113.

Prihar, S.S., Singh, R., Singh, N. and Sandhu, K.S. (1979) Effects of mulching previous crops or fallow on dryland maize and wheat. *Exp. Agric.* 15, 129–134.

Rees, R.G., and Platz, G.J. (1979) The occurrence and control of yellow spot of wheat in north-eastern Australia. *Aust. J. Exp. Agric. Anim. Husb.* 19, 369–372.

Reeves, T.G. and Brooke, H.D. (1977) The effect of genotype and phenotype on the competition between wheat and annual ryegrass. Proc. 6th Asian Pacific Weed Science Conf., Jakarta, 1977, pp. 166–171.

Reeves, T.G. and Ellington, A. (1974) Direct drilling experiments with wheat. *Aust. J. Exp. Agric. Anim. Husb.* 14, 237–240.

Rowell, D.L., Osborne, G.J., Matthews, P.G., Stonebridge, W.C. and McNeill, A.A. (1977) The effects in a long-term trial of minimum and reduced cultivation on wheat yields. *Aust. J. Exp. Agric. Anim. Husb.* 17, 802–811.

Salinas, J.G. and Sanchez, P.A. (1976) Soil plant relationships affecting varietal and species differences in tolerance to low available soil phosphorus. *Cienc. Cult.* 28, 156–168.

Souza Machado, V., Bandeen, J.D., Stephenson, G.R. and Lavigne, P. (1978) Uniparental inheritance of chloroplast atrazine tolerance in *Brassica campestris. Can. J. Plant Sci.* 58, 977–981.

Swarbrick, J.T. (1980) Weed management in Australian rural production, 1980–1990. Proc. Aust. Agron. Conf., Lawes, 1980. pp. 146–156.

Syme, J.R., Mackenzie, J. and Strong, W.M. (1976) Comparison of four wheat cultivars for yield and protein response to nitrogen fertilizer. *Aust. J. Exp. Agric Anim. Husb.* 16, 407–410.

Unger, P.W. (1978) Straw-mulch rate effect on soil water storage and sorghum yield. *Soil Sci. Soc. Am. J.* 42, 486–491.

Van Doren, D.M. Jr. and Allmaras, R.R. (1978) Effect of residue management practices on the soil physical environment, microclimate, and plant growth. In 'Crop Residue Management Systems'. (Ed. W.R. Oschwald.) American Society of Agronomy Special Publication No. 31, pp. 49–83.

16 PROSPECTS

R. Riley

For almost 60 years Otto Frankel has participated in and watched the unfolding of wheat science. He was born on the 4th November 1900 the year in which, on the 4th May, William Bateson, on his way to London from Cambridge to lecture to the Royal Horticulture Society, read Gregor Mendel's paper for the first time and immediately modified his lecture to refer to Mendel's findings. As a young scientist Otto Frankel worked during the maximum flowering of 'non-molecular genetics', a time when the intellectual demands were even greater than at present, now that we can describe genetic phenomena in terms of chemistry. He observed and contributed to the administration of science over the period that biology became molecular, and increasingly he has contributed thought and action to the conservation of genetic resources. He has added to knowledge by scientific research and to the intellectual tenor of our age by the keenness and rigour of his thoughts, which he has always communicated with great clarity.

Wheat production

The progress in wheat production that I have been able to watch during the 28 years that I have worked on the crop in the United Kingdom has been remarkable. Figure 16.1, for which I am indebted to R.B. Austin of the Plant Breeding Institute, shows that the national mean yield has progressed from approximately 2.5 tonnes per hectare in the immediate post-war years to about 5.65 tonnes in 1980. Over this period the mean annual increase has been 0.076 ± 0.005 t/ha. The attainment of this increased productivity has depended upon many factors including, prominently, economic stimuli and incentives provided by government, a progressive restructuring of the industry, a perceptive farming community and a skilled and adaptable work force. However, it could not have been achieved without the vigorous application of science to the entire range of farm practice from primary cultivation and seed bed preparation, the application of fertilizers in correct balance, timing and placement, effective weed control and crop protection to timely and speedy harvesting, drying and storage of the grain.

I have not mentioned, so far, the consequence of variety on the gain in yield in the U.K. but recent studies have shown genetic advance to have been of very considerable significance. Silvey (1978) has calculated, by comparing the

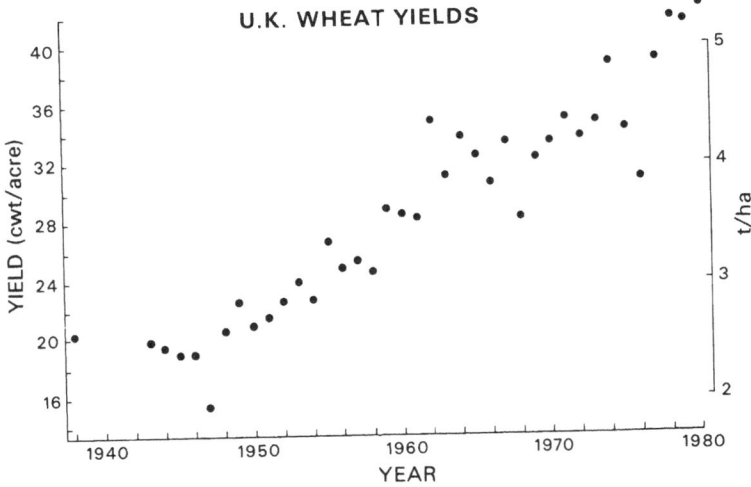

Fig. 16.1. Mean yield of wheat in the United Kingdom.

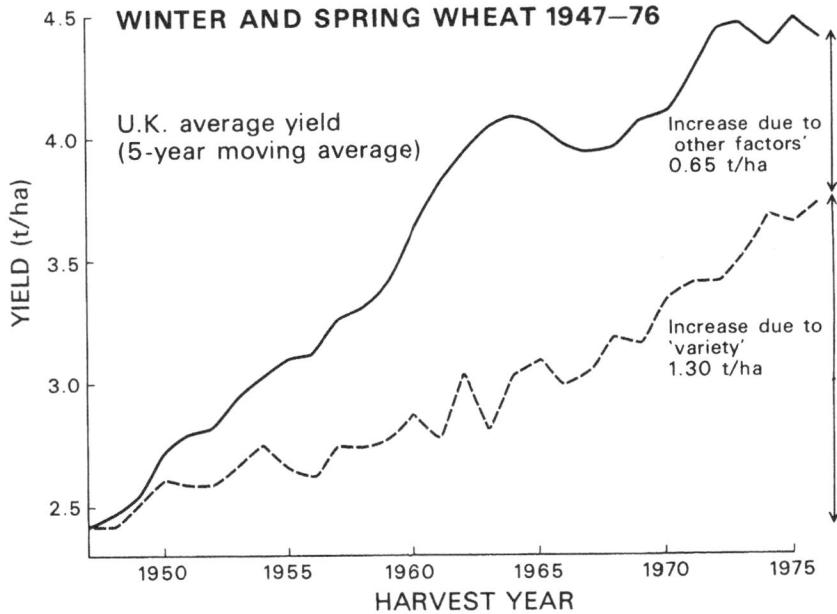

Fig. 16.2. Mean yield of wheat in the United Kingdom expressed as a moving 5-year average and compared with the estimated contributions to the annual increase in yield due to variety. (Silvey, 1978).

proportion of the crop area occupied by each succeeding variety and the relative yields of each variety with the advancing annual mean yield computed on a rolling five year bais, that between 1947 and 1976, 56 per cent of the 2.04 t/ha increase in mean yield was due to variety (cf. Figure 16.2).

My colleagues (Austin *et al.*, 1980) have compared older and new winter wheat varieties in trials. This has been made possible by availability of broad spectrum fungicides which can protect the older varieties from currently prevalent races of pathogens while the disadvantage of weaker straw in the older varieties can be obviated by supporting the entire trial by netting. On this basis the variety Norman, released in 1980, was shown to yield about 50% more than such pre-war varieties as Little Joss and Holdfast. The organization of the trial discounted the benefits from improved straw strength and disease resistance (particularly to eyespot) in modern varieties and when allowance is made for these factors, the increase in yield ascribed to variety from this experiment is rather similar to that estimated by Silvey (1978). The gain in grain yield has been achieved very largely by changes in the harvest index from about 35% in the older varieties to about 50% in Norman. This is a remarkable achievement: the contribution of breeding to U.K. wheat production has added £130 return per hectare (at present prices), or almost £170M annually, to the value of the national crop. However, after improving the partitioning of dry matter there is now a bigger challenge facing plant physiologists and breeders, namely that of improving overall production efficiency. Plant physiology must show how photosynthetic efficiency can be increased per unit area of leaf, or how the total dry matter production can be increased without reducing harvest index.

So far I have discussed the yield potential of the wheat crop but, of course, in many regions prime importance may rest not on the need to increase the yield potential of the crop but on understanding why mean yields do not approximate more closely the existing potential. Thus in the United Kingdom the potential yield of winter wheat has been calculated by R.B. Austin (personal communication) to be about 12.6 t/ha (Table 16.1).

In the 1980 harvest a farmer in Suffolk in eastern England, without excessive inputs except of skill, produced approximately 11.55 t/ha from a farm scale crop of more than 200 acres of the variety Hustler. So yield not far short of the theoretical maximum can be harvested. Yet in the same year the national average yield when finally computed will probably be about 5.65 t/ha and, excellent though this is, an understanding of the causes of the shortfall between mean and maximum yields could lead to the better deployment of existing technology with useful economic returns. Of course, some of the constraints to yield advance are undoubtedly economic and social and some are technical but

Table 16.1

The potential yield of wheat in England

	t/ha	
Photosynthesis		
Leaves and stems	14.4	
Ears	2.6	
	17.0	
Respiration, 40% of above	6.8	
Net photosynthesis		10.2
Pre-anthesis assimilation		0.5
Total grain drymatter		10.7
Moisture to give 15% fresh weight basis		1.9
Potential yield		12.6

(Assumptions: 40 day grain filling period, P max = 30 mg $CO_2/dm^2/h$, LAI at anthesis = 10, Solar radiation = 16.8MJ/m^2/day.)

not amenable to modification, like the basic physical structure of the soil. But many constraints, if precisely identified, could be changed.

It was recognition of this problem that stimulated the Agricultural Research Council to initiate a priority programme to study yield variation in winter wheat. There are several components to this programme, but the most complex is that at the Rothamsted Experimental Station (Lester, 1980). An annual series of multifactorial experiments has been initiated in which eight factors each at two levels are being studied in a half replicated design of 128 plots. Additional auxiliary experiments are being used to test N response and to sample a contrasted soil type. Many of the Station's departments are working on the experiments, which for the 1978/79 season emphasized the importance of controlling leaf diseases and aphids in June and July (grain filling). Early sowing reinforced the benefits of fungicides and aphicide applications. It is anticipated that greater rigour will be applied to commercial crop management when the results of series of experiments of this kind have been fully assessed. I look for future progress, in practical wheat science, to the employment of truly multidisciplinary studies and our science arm must be organized to encourage such multidisciplinary research.

Wheat genetics

Understanding of the structure of the genomes of wheat and its relatives has advanced remarkably over the past three decades. At the outset, we were beginning, due to the brilliance of E.R. Sears and his collaborators, to understand the nature of homoeology, while now we are beginning to learn about the nucleotide structure of the DNA in particular chromosome regions like the six different, highly repeated, sequences that occur in the heterochromatic telomeres of rye, four of which are specific to *S. cereale* (Bedbrook *et al.*, 1980). A major problem still remains in understanding the function of much of the wheat chromosome DNA content because of the following:

— The estimated DNA content of a 4C nucleus of *T. aestivum* is 69.3 pg (Bennett and Smith 1976). If we wish to assess what proportion of this may represent informational content it is necessary to consider only one of the three constituent genomes A, B and D.

— Assuming that the DNA of *T. aestivum* is equally divided among the three genomes, the 4C DNA content of a genome is 23.1 pg.

— However, we are concerned with the irreducible DNA content of a genome, where each locus is present in single dose and this is given by the DNA content of a 1C nucleus of a genome which is 5.8 pg.

— 5.8 pg of DNA approximates to 5.4×10^9 nucleotide pairs.

— If it is assumed that the mean protein size is about 330 amino acid residues, then the mean gene length is about 10^3 base pairs.

— If it is further assumed that wheat has about 10^4 genes in each genome (a number held to be appropriate for many diploid organisms), then 10^7 nucleotide pairs are components of those genes.

— As a percentage of the total genome, these 10^7 nucleotide pairs would represent 0.185%.

Clearly these calculations make many assumptions and are imprecise. Their significance is nevertheless clear in indicating that a very small percentage of wheat DNA — probably considerably less than one per cent — codes for protein production. Much of the work of plant molecular biologists over the coming decade will lie in describing and explaining the organizational and functional properties of the remaining DNA.

This will be a necessary preliminary to attempts to introduce new genetic information into wheat by recombinant DNA technology. However, work with this purpose is already launched. In the United Kingdom, contributions to the total programme are being made from a number of specialist laboratories so that convergence of knowledge about several distinct but essential technologies can occur. New potentialities for the transformation of plants has recently been

provided by the exciting discovery of Clarke and Carbon (1980) that the incorporation of the DNA of a centromere into a yeast plasmid causes it to behave
essentially like a chromosome at meiosis. Clearly this process may enable the
development of new schemes for the genetic manipulation of plants including
wheat.

Grain quality

Although much attention has been devoted over the years to the means of
breeding for breadmaking quality, until recently no clear cut relationship had
been detected between a simple grain characteristic and quality. However,
Payne *et al.* (1979) have now shown that, when the composition of glutenin of
U.K. wheats was examined by SDS-polyacrylamide-gel electrophoresis, the
presence of a particular subunit (approximate molecular weight of 145 000),
determined by chromosome 1A, was strongly associated with breadmaking
quality. Although the presence of this subunit is not essential for good bread-
making quality and it is present in some genotypes which have no bread-making
capabilities, nevertheless within the west-European collection of varieties there is
a very strong association between its presence in a variety and good quality.
These results give hope that greater precision may be introduced into selection
for bread-making quality in the future.

Disease resistance

We have learnt in the U.K. that no reliance can be placed on the protection of
the wheat crop by the use of varieties whose resistance to stripe rust is determined by major race-specific genes. Increasingly we are attempting to use what is
termed 'durable' resistance. This is a form of inherited resistance which may
not give total protection but which shields the crop from major epidemics for
several years when the variety containing it occupies a considerable proportion
of the crop area. The winter wheat variety Bersee behaved in this way and
Johnson and Law (1975) demonstrated that its durable resistance is determined
by activities of chromosome $5B^S/7B^S$. The resistance is compounded of two or
more separate activities of the chromosome but because the chromosome
can be substituted intact in breeding its separate elements can be held together.
As we look to the future the prospects of handling disease resistance of this
complex nature must offer promise of improved crop protection.

Conclusions

In this paper I have made reference to a few sectors of wheat science where I
expect useful progress to be made in the future and where, indeed, progress is
necessary. Of course, wheat science exists because of the dependence of man on

the wheat plant, and research which was initiated for solely practical purposes has proliferated to provide a wide range of knowledge, the availability of which makes wheat the organism of choice for much research with no pretensions to utility. However, the connections between fundamental and applied aspects will always be strong in wheat science. The knowledge derived is essential for the support of the wheat industry in both its production and its processing phases. This symposium has demonstrated that wheat science is creative and abundantly productive and that the flow of new ideas into practice continues unabated. There is, therefore, every prospect that wheat technology will continue to develop along existing and along new lines to the benefit of mankind.

The most effective benefits, both scientific and practical, will increasingly come in the future from multidisciplinary research of the kind that I have mentioned is under way at Rothamsted. As part of this development, we shall also need the ability to synthesize the many different themes that have made up this symposium so that there will genuinely be a unitary science of wheat. This symposium will have achieved a great deal if it shows a way forward to this goal, as well as by enabling us to pay tribute to Otto Frankel by assessing the status of 'Wheat Science — Today and Tomorrow'.

References

Austin, R.B., Bingham, J., Blackwell, R.D., Evans, L.T., Ford, M.A., Morgan, C.L. and Taylor, M. (1980) Genetic improvements in winter wheat yields since 1900 and associated physiological changes. *J. Agric. Sci.* **94**, 675–689.

Bedbrook, J.R., Jones, J., O'Dell, M., Thompson, R.D. and Flavell, R.B. (1980) A molecular description of telomeric heterochromatin in *Secale* species. *Cell* **19**, 545–560.

Bennett, M.D., and Smith, J.B. (1976) Nuclear DNA amounts in angiosperms. *Philos. Trans. R. Soc. Lond. B Biol. Sci.* **274**, 227–274.

Clarke, L., and Carbon, J. (1980). Isolation of a yeast centromere and construction of functional small circular chromosomes. *Nature* **287**, 504–509.

Johnson, R., and Law, C.N. (1975) Genetic control of durable resistance to yellow rust (*Puccinia striiformis*) in the wheat cultivar Hybride de Bersee. *Ann. Appl. Biol.* **81**, 385–391.

Lester, E. (1980). Multidisciplinary activities. *Rothamsted Exp. Stn. Rep. 1979*, 17-25.

Payne, P.I., Corfield, H.G., and Blackman, J.A. (1979) Identification of a high-molecular-weight subunit of glutenin whose presence correlates with bread-making quality in wheat of related pedigree. *Theor. Appl. Genet.* **55**, 153–159.

Silvey, V. (1978) The contribution of new varieties to increasing cereal yield in England and Wales. *J. Natl. Inst. Agric. Bot.* **14**, 367–384.

17 OTTO FRANKEL – CONTRIBUTIONS TO WHEAT GENETICS

M.J.D. White

The contributions of Otto Frankel to the advancement of science over the past 55 years range from intellectually and technically sophisticated investigations of the genetic basis of developmental systems in the wheat plant to popularization – in the best sense of the word – of the need to conserve genetic resources in a world in which the basic foundations of agriculture are constantly in flux.

Otto Frankel's work on the wheat plant falls into three very distinct categories. First comes his long series of investigations on yield and quality in New Zealand wheats (1930, 1932, 1934, 1935, 1939, 1940, 1941, 1947, 1948; Frankel and Donald, 1933; Boyce, Copp and Frankel, 1947). Concerning these I shall only say that they were obviously carried out with extreme methodological rigour and seem to have been eminently successful from a practical standpoint. Secondly, there is a detailed cytogenetic study of inverted duplications in wheat (1949). The influence of the John Innes Horticultural Institution and especially of its leading spirit, C. D. Darlington, which is evident in Frankel's early cytogenetic work on liliaceous plants, is also manifest in his complex analysis of these terminal inverted duplications which occurred in an F_5 progeny derived from a cross between the varieties Tuscan and White Fife. Two such inverted duplications, a long and a short one, were studied. They were interpreted as having arisen by breakage of dicentric chromatid bridges but this requires that the broken end should undergo 'healing', i.e. that it should convert itself into a functional telomere, so that no sister reunion of chromatids takes place. Apparently this kind of healing process, following breakage of meiotic bridges, is common in wheat (unlike the situation in maize where, as shown by McClintock, it does not occur at meiosis, although it can occur at mitosis).

Frankel obtained homozygotes and heterozygotes for these duplications and also the combined long/short heterozygotes. The meiotic behaviour was studied in detail. Internal chiasma formation (within the chromosome arms) was seen in all genotypes. Chiasma formation between chromosomes was strongly reduced or absent in duplication/normal heterozygotes but occurred in long/short heterozygotes and in duplication homozygotes. Inversion crossing over gradually removes the duplications from the population; thus if a heterozygote for the long duplication (LD/N) is bred without selection, the long duplication was calculated to have an existence limited to 13 generations. Frankel draws some analogies between inverted terminal duplications and supernumerary chromosomes, but these do not seem particularly relevant today. Nevertheless, this investigation stands as a piece of classic cytogenetics, important because of the differences in behaviour between wheat and other organisms, especially maize.

In the same material as that in which the duplications occurred, a mutant chlorophyll defect was found (Frankel, 1950) which led to somatic variegation (leaves with green and white sectors). This *striato-virescens* condition was shown to be due to three recessive mutations, presumably located on homoeologous chromosomes of the A, B and D genomes. The origin of this condition is obscure. Chlorophyll mutants are generally uncommon in cereals, even in diploid species. It was established that neither of the original parental plants carried any of the mutant recessives. Frankel considered two possibilities: (i) that one gene mutated to the recessive allele and was transferred to the two homoeologous chromosomes, presumably by crossing over, and (ii) that a multiple mutation occurred. He opts for the second alternative, on the grounds that there was insufficient time between the parental generation and the F_5 one for three normally rare events to occur. It would be interesting to know whether any evidence has accumulated in the past thirty years bearing on this question - the writer is ignorant on this point.

Undoubtedly, Frankel's most important contribution to wheat genetics has been his long series of studies on basal sterility (i.e. sterility of the lower florets of the spikelets) in speltoid wheats (Frankel and Fraser, 1948; Frankel, Gani and Munday, 1958; Frankel, Shineberg and Munday, 1969; Frankel and Roskams, 1975; Frankel, 1975, 1976). This is one of the most thoroughly investigated examples of organ morphogenesis in plants.

So-called speltoid wheats have a brittle main axis of the spike and tough glumes; when threshed the axis breaks and the grains remain within the glumes. 'Naked' or *vulgare*-type wheats have a tough axis and loose glumes so that when threshed the axis remains unbroken and the grains are readily detached. The spelt condition is clearly primitive, since it is present in the diploid species of *Triticum* and in related diploid species of *Aegilops*. The 'naked' character makes its first appearance in some of the tetraploid wheats (*T. durum, T. turgidum, T. polonicum, T. turanicum* and *T. carthlicum*). Among the hexaploid wheats the subspecies *spelta, macha* and *vavilovi* are spelt, *vulgare, compactum* and *sphaerococcum* are naked, due to the presence of the so-called Q factor, or segment, on the long arm of chromosome 5 of the A genome, derived from the diploid *T. boeoticum*. In the *vulgare*-type wheats, speltoid mutants arise from time to time and are apparently always due to deletions of the Q segment, because back-mutation never takes place.

Frankel's studies were concerned with the morphogenetic action of a gene Bs,bs, that plays a major role in the control of floral development in the absence of Q. The Bs gene acts as a weaker Q, inducing fertility only from the second flower onward, whereas Q acts from the first flower onward. The presence of Bs is compatible with a range of expression, from full fertility to complete absence of the first flower, depending on gene dosage ($BsBs$ or $Bsbs$) and on polygenic background. Frankel (1976) studied the effects of such environmental variables as day length and temperature on floral morphogenesis in different wheat genotypes. For each floret from the base to the apex there is a sensitive period intervening between the initiation of the subtending bract or lemma and that of the floret itself. Initiation of spikelet development begins in the central region of the inflorescence and spreads in both directions. Frankel calls this the *Developmental gradient*. It is similar in normal wheats and speltoids with basal sterility and is not much affected by environmental treatments. The *basal sterility gradient* typically (in speltoids St_2 and St_3) produces a range of phenotypes from relatively normal florets in spikelets near the apex to abnormal ones in spikelets near the base.

The difference between speltoid stocks indicates that the basal sterility gradient is under genetic control. Frankel has postulated that there is an activator substance that interacts with the normal developmental gradient and is itself under a system of genetic control independent of the system that governs the developmental pattern of the inflorescence.

In earlier work, Frankel, Shineberg and Munday (1969) studied the genetic basis of canalization (to use Waddington's term) for basal fertility in the wheat inflorescence. It is clear that in all this work Frankel was strongly influenced by the ideas of the animal geneticists Waddington, Rendel and his own student A. S. Fraser. On the other hand he has always been an independent worker, with able colleagues supporting him. The idea that environmental effects on basal sterility should be systematically investigated was partly suggested by the observation of marked fluctuations in the degree of basal sterility between seasons and partly by the discovery by Evans of environmental effects on the development of the inflorescence in various grasses.

Frankel used his speltoid tester stocks to examine the genetic basis of basal fertility in subspecies of *Triticum*, both tetraploid and hexaploid, that do not have Q. He found the control to be the same polygenic system as in *T. aestivum*. Since basal sterility, although uncommon in the tribe Hordeae, to which the wheats belong, is found sporadically throughout the Gramineae, the genetic and developmental mechanisms responsible for its manifestation are likely to be of considerable evolutionary antiquity.

The opposite of basal sterility is, in one sense, the conversion of the normally sterile glumes into flowering organs. Frankel (1976) suggests that basal sterility and glume fertility are the opposite ends of a developmental gradient for floral morphogenesis, and that this gradient is controlled by the genetic system first made manifest in base-sterile speltoids. A high dosage of fertility-stimulating polygenes in a genotype bearing Q, or an overdose of Q in a plant tetrasomic for it, causes glumes to develop flowers.

Clearly, Frankel's analysis of basal sterility in wheat deserves to be carried forward using biochemical methods and with biochemical ideas. Identification of the activator substance, or substances, that interact with the developmental gradient would be an important step forward. Future studies are likely to involve *in vitro* organ culture and biochemical modification of floral morphogenesis. Frankel's wheat research is open-ended in the sense that he has pointed the way rather than provided final answers. In most respects the study of developmental systems in animals has progressed further than that of plant systems. But a few systems stand out as examples of what can be done with plant material and basal sterility in wheat is emphatically one of these.

A detailed account of Otto Frankel's contribution to the conservation of genetic resources would be out of place here, but some mention must be made of his philosophical approach, particularly as applied to crop plants, including wheat. One of his most important publications in this field is undoubtedly the paper he presented at the Thirteenth International Congress of Genetics in Berkeley (Frankel, 1974). This will undoubtedly be regarded in years to come as a landmark in the cultural evolution of the human species. Here he defines the 'time scale of concern' for domesticated plants and animals and for wild species. The social, ecological and genetic parameters of conservation are considered in respect of natural communities, artificial ones (zoological and botanical gardens) and domesticated plants. Frankel's approach is not simply utilitarian or humanitarian (feeding the world's human population) but embraces the whole range of the cultural significance of the earth's biota. His view is a somewhat pessimistic but certainly realistic one. As far as economic plants are concerned he concludes that the most important genetic resources that need to be conserved are the primitive varieties of traditional indigenous agriculture now threatened with extinction by modern cultivars. Wild species related to cultivated plants are, in general, less threatened at the present time.

With regard to wild species of animals and plants, Frankel has been especially interested in the evidence for minimal viable population sizes as emphasised by the studies of Michael Soulé on lizards and of Main and Yadov on marsupials, both carried out on island populations. This aspect will be emphasised in a forthcoming book, written in collaboration with Soulé. I, personally, am more sceptical than Frankel as to the reliability of estimates based on these studies and the general validity of minimal populaton sizes based on genetic considerations. However, this is a detail. In his general approach, Frankel has established himself as a worthy successor to Vavilov, and if he has not emulated the field collecting of the great Russian plant geneticist, his intellectual contribution to the problems of the conservation of biological resources has been even wider and is likely to have a lasting impact in human history. His work in organising the International Genetics Federation has played a large part in bringing together pure and applied geneticists in order to confront these most critical problems of the earth's biota in an intelligent, informed and humane manner.

PUBLICATIONS OF O.H. FRANKEL

Frankel, O.H. (1925) Faktorenkoppelung bei Pflanzen. *Z. Indukt. Abstammungs Vererbungsl.* **38**, 324–348.

Oppenheim, J.D. and Frankel, O.H. (1929) Investigations into the fertilization of the Jaffa orange. I. *Genetica* **11**, 369–374.

Frankel, O.H. (1930) Genetics and plant breeding. *N.Z. J. Sci. Technol.* **11**, 401–408.

Frankel, O.H. (1930) Analytical yield investigations of New Zealand wheat. 1. Wheat Res. Inst. 1st Annu. Rep. pp. 42–59.

Frankel, O.H. (1932) Analytische Ertragsstudien an Getreide. *Der Zuechter* **4**, 98–109.

Frankel, O.H. and Donald, H.P. (1933) Some critical observations on quality testing in wheat breeding. Proc. Conf. World Grain and Exhibition, 2, pp. 400–408.

Frankel, O.H. (1933) A case of mass-occurrence of non-inherited chlorophyll defects in wheat. *Trans. R. N.Z. Inst.* **63**, 141–143.

Frankel, O.H. (1934) 'Cross 7 wheat.' A new combination of high yield and baking-quality. *N.Z. Dep. Sci. Ind. Res. Bull.* **46**.

Frankel, O.H. (1935) The differentiation of grain samples of closely related varieties of wheat by means of a simple mechanical test for grain quality. *J. Agric. Sci.* **25**, 461–465.

Frankel, O.H. (1935) Analytical yield investigations on New Zealand wheat. 2. Five years analytical trials. *J. Agric. Sci.* **25**, 466–509.

Frankel, O.H. (1937) Inversions in *Fritillaria*. *J. Genet.* **34**, 447–462.

Frankel, O.H. and Hair, J.B. (1937) Studies on the cytology, genetics and taxonomy of N.Z. *Hebe* and *Veronica*. Part 1. *N.Z. J. Sci. Technol.* **18**, 669–687.

Frankel, O.H. and Hair, J.B. (1938) Analytical yield investigations on N.Z. wheat. III. Nine years' observations on two varieties. *N.Z. J. Sci. Technol.* **20A**, 244–259.

Frankel, O.H. (1939) Tainui, a new spring wheat variety. *N.Z. J. Sci. Technol.* **20A**, 319–323.

Frankel, O.H. (1939) Analytical yield investigations on N.Z. wheat. 4. Blending varieties of wheat. *J. Agric. Sci.* **29**, 249–261.

Frankel, O.H. (1940) A critical survey of breeding wheat for baking quality. *J. Agric. Sci.* **30**, 98–112.

Frankel, O.H. (1940) Studies in *Hebe*. II. The significance of male sterility in the genetic system. *J. Genet.* **40**, 171–184.

Frankel, O.H. (1941) The causal sequence of meiosis. I. Chiasma formation and the order of pairing in *Fritillaria. J. Genet.* **41**, 9–34.

Frankel, O.H. (1941) Cytology and taxonomy of *Hebe, Veronica* and *Pygmaea. Nature (Lond.)* **147**, 117–118.

Frankel, O.H. (1941) 'Fife-Tuscan' wheat: a new variety for 'Tuscan land'. *N.Z. J. Sci. Technol.* **22A**, 303–308.

Frankel, O.H. (1947) Plant collections. *J. Aust. Inst. Agric. Sci.* **13**, 122–124.

Boyce, S.W., Copp, L.G.L. and Frankel, O.H. (1947) The effects of selection for yield in wheat. *Heredity* **1**, 223–233.

Frankel, O.H. (1947) The theory of plant breeding for yield. *Heredity* **1**, 109–120.

Frankel, O.H. (1948) Hilgendorf wheat of outstanding baking quality. *N.Z. J. Agric.* **76**, 117–119.

Frankel, O.H. (1948) A new high-yielding wheat variety — WRI yielder. *N.Z. J. Agric.* **76**, 221–222.

Frankel, O.H. and Fraser, A.S. (1948) Basal sterile mutants in speltoid wheat. *Heredity* **2**, 391–397.

Frankel, O.H. (1949) A self-propagating structural change in *Triticum*. I. Duplication and crossing-over. *Heredity* **3**, 163–194.

Frankel, O.H. (1949) A self-propagating structural change in *Triticum*. II. The reproductive cycle. *Heredity* **3**, 293–317.

Frankel, O.H. (1950) The development and maintenance of superior genetic stocks. *Heredity* **4**, 89–102.

Frankel, O.H. (1950) A polymeric multiple gene change in hexaploid wheat. *Heredity* **4**, 103–116.

Frankel, O.H. (1951) The multiple mutation in wheat. *Heredity* **5**, 349.

Frankel, O.H. (1954) Invasion and evolution of plants in Australia and New Zealand. *Caryologia* **6**, 600–619.

Frankel, O.H. (1954) Genetic adaptation of cultivated plants in Australia. Proc. Pan Indian Ocean Sci. Congr., Perth, W.A. pp. 71–74.

Frankel, O.H. (1957) The biological system of plant introduction. *J. Aust. Inst. Agric. Sci.* **23**, 302–307.

Frankel, O.H. (1958) The biological system of plant introduction. *Indian J. Genet. Plant Breed.* **17**, 336–342.

Frankel, O.H. (1958) The dynamics of plant breeding. *J. Aust. Inst. Agric. Sci.* **24**, 112–123.

Frankel, O.H. and Williams, J.D. (1958) A record of natural crossing in subterranean clover. *J. Aust. Inst. Agric. Sci.* **24**, 162–163.

Frankel, O.H., Gani, R. and Munday, A. (1958) Two independent gene systems for floral induction in wheat. *Proc. 10th Int. Congr. Genet.* **2**, 86.

Frankel, O.H. (1959) Variation under domestication. *Aust. J. Sci.* 22, 27–32.

Morley, F.H.W. and Frankel, O.H. (1959) An ecogenetic research programme with introduced plants. 'Biogeography and Ecology in Australia'. *Monogr. Biol.* 8, 577–586.

Brock, R.D. and Frankel, O.H. (1960) Plant improvement. *J. Aust. Inst. Agric. Sci.* 26, 170–182.

Frankel, O.H. and Munday, A.M. (1960) The genetics of floral development in wheat. J. Wheat Inform. Serv. Biol. Lab. Kyoto Univ. Kyoto, Japan. No. 11, 1–2.

Frankel, O.H. (1961) The FAO Freedom From Hunger Campaign. *J. Aust. Inst. Agric. Sci.* 27, 79–84.

Frankel, O.H. and Munday, A. (1962) The evolution of wheat. In 'The Evolution of Living Organisms'. R. Soc. Victoria Darwin Cent. Symp., Melbourne, 1959. pp. 173–180.

Frankel, O.H. (1962) Agricultural science and productivity in the next decade: plant science. *J. Aust. Inst. Agric. Sci.* 28, 84–91.

Frankel, O.H. (1963) Concluding remarks: The next decade. In 'Environmental Control of Plant Growth'. Academic Press Inc., New York. pp. 439–441.

Frankel, O.H. (1963) Agricultural scientists among scientists. *J. Aust. Inst. Agric. Sci.* 29, 95–103.

Frankel, O.H. and Munday, A. (1963) Canalization of flower morphogenesis in wheat. *Proc. 11th Int. Congr. Genet.* 1, 10–31.

Frankel, O.H. (1963) The social responsibility of agricultural science. The Farrer Memorial Oration, 1962. *Aust. J. Sci.* 25, 301–307.

Barnard, C. and Frankel, O.H. (1964) Grass, grazing animals and man in historic perspective. In 'Grasses and Grasslands'. (Ed. C. Barnard.) MacMillan & Co., London. pp. 1–12.

Frankel, O.H. (1966) Internationalism in agricultural science. *Aust. J. Sci.* 28, 314–320.

Frankel, O.H. (1966) The International Biological Program. *Aust. J. Sci.* 28, 324–325.

Frankel, O.H. (1966) Adaptability of crops. *New Sci.* 31, 144–145.

Frankel, O.H. (1967) Guarding the plant breeder's treasury. *New Sci.* 35, 538–540.

Frankel, O.H. (1968) International collaboration in plant exploration and conservation. *J. Aust. Inst. Agric. Sci.* 34, 22–27.

Frankel, O.H. (1968) Human welfare and international co-operation. *Proc. Natl. Acad. Sci. U.S.A.* 60, 33–41.

Frankel, O.H. (1968) Man in the biosphere: An international study. Symp. Int. Biol. Prog. Aust. Acad. Sci. 1967. Union Offset Co., Canberra, A.C.T. pp. 4–12.

Frankel, O.H. and Shineberg, B. (1969) The genetic system of basal fertility in wheat. Proc. 3rd Int. Wheat Genetics Symp., Canberra, 1968. pp. 279–281.

Frankel, O.H. (1969) The dynamics of plant breeding. *Proc. 12th Int. Congr. Genet.* 3, 309–325.

Frankel, O.H., Shineberg, B. and Munday, A. (1969) The genetic basis of an invariant character in wheat. *Heredity* 24, 571–591.

Frankel, O.H. (1969) Pacific centres of genetic diversity. *Malay. For.* 32, 356–360.

Frankel, O.H. (1970) Save the genetic treasuries in the Sabrao region. *Sabrao Newsl.* 2, 1–6.

Frankel, O.H. (1970) Genetic conservation of plants useful to man. *Biol. Conserv.* 2, 162–169.

Frankel, O.H. (1970) Genetic dangers in the green revolution. *World Agric.* 19, 9–13.

Frankel, O.H. (1970) Preface. In 'Genetic Resources in Plants – their Exploration and Conservation'. IBP Handbook No.11. (Eds. Frankel, O.H. and Bennett, E.) Blackwell, Oxford and Edinburgh. pp. 1–4.

Frankel. O.H. and Bennett, E. (1970) Genetic resources. In 'Genetic Resources in Plants – their Exploration and Conservation'. IBP Handbook No.11. (Eds. Frankel, O.H. and Bennett, E.) Blackwell, Oxford and Edinburgh. pp. 7–17.

Frankel, O.H. (1970) Evaluation and utilization – Introductory remarks. 'Genetic Resources in Plants – their Exploration and Conservation'. IBP Handbook No.11. (Eds. Frankel, O.H. and Bennett, E.) Blackwell, Oxford and Edinburgh. pp. 395–401.

Frankel, O.H. (1970) Genetic conservation in perspective. In 'Genetic Resources in Plants – their Exploration and Conservation'. IBP Handbook No.11. (Eds. Frankel, O.H. and Bennett, E.) Blackwell, Oxford and Edinburgh. pp. 469–489.

Frankel, O.H. and Bennett, E. (Eds.) (1970) 'Genetic Resources in Plants – their Exploration and Conservation'. IBP Handbook No.11. Blackwell, Oxford and Edinburgh. 554 pp.

Frankel, O.H. (1970) Variation – the essence of life. *Proc. Linn. Soc. N.S.W.* 95, 158–169.

Frankel, O.H. (1972) Australia and the International Biological Program. *Search (Syd.)* 3, 105–108.

Frankel, O.H. (1972) Genetic conservation – a parable of the scientist's social responsibility. *Search (Syd.)* 3, 193–201.

Frankel, O.H. (1972) Only One Earth: The United Nations Conference on the Human Environment, Stockholm, 5–16 June 1972. *Search (Syd.)* 3, 406–408.

Frankel, O.H. (Ed.) (1973) 'Survey of Crop Genetic Resources in their Centres of Diversity'. FAO/IBP, Rome. 164 pp.

Frankel, O.H. (1973) Introduction. In 'Survey of Crop Genetic Resources in their Centres of Diversity'. (Ed. Frankel, O.H.) FAO/IBP, Rome, pp. ix—xiv.

Frankel, O.H. (1974) Genetic conservation: our evolutionary responsibility. *Genetics* **78**, 53—65.

Frankel, O.H. and Roskams, M. (1975) Stability of floral differentiation in *Triticum. Proc. R. Soc. Lond. B. Biol. Sci.* **188**, 139—162.

Frankel, O.H. (1975) Base-sterile speltoids: the location of the *Bs* gene of *Triticum aestivum. Proc. R. Soc. Lond. B Biol. Sci.* **188**, 163—166.

Frankel, O.H. and Hawkes, J.G. (Eds.) (1975) 'Crop Genetic Resources for Today and Tomorrow.' (International Biological Programme 2). Cambridge University Press, London. 492 pp.

Frankel, O.H. and Hawkes, J.G. (1975) Genetic resources – the past ten years and the next. In 'Crop Genetic Resources for Today and Tomorrow'. (International Biological Programme 2). (Eds. Frankel, O.H. and Hawkes, J.G.) Cambridge University Press, London. pp. 1—11.

Frankel, O.H. (1975) Genetic resources centres – a co-operative global network. In 'Crop Genetic Resources for Today and Tomorrow'. (International Biological Programme, 2). (Eds. Frankel, O.H. and Hawkes, J.G.) Cambridge University Press, London. pp. 473—481.

Frankel, O.H. (1975) Genetic resources survey as a basis for exploration. In 'Crop Genetic Resources for Today and Tomorrow'. (International Biological Programme, 2). (Eds. Frankel, O.H. and Hawkes, J.G.) Cambridge University Press, London. pp. 99—109.

Frankel, O.H. (1975) Conservation in perpetuity: ecological and biosphere reserves. In 'A National System of Ecological Reserves in Australia'. (Ed. Fenner, F.) Australian Academy of Science, Canberra. pp. 7—10.

Frankel, O.H. (1975) Genetic conservation – why and how. In 'South East Asian Plant Genetic Resources'. (Eds. Williams, J.T., Lamoureux, C.H. and Wulijarni-Soetjipto, N.) Proc. Symp. Bogor, 1975. International Board for Plant Genetic Resources and others, Bogor, Indonesia. pp. 16—32.

Frankel, O.H. (1976) Floral initiation in wheat. *Proc. R. Soc. Lond. B Biol. Sci.* **192**, 273—298.

Frankel, O.H. (1976) The time scale of concern. In 'Conservation of Threatened Plants'. (Eds. Simmons, J.B., Beyer, R.I., Brandham, P.E., Lucas, G.Ll. and Parry, V.T.H.) Plenum Press, New York. pp. 245—248.

Frankel, O.H. (1976) Biological structure of the landscape. In 'Man and Landscape in Australia: towards an Ecological Vision'. (Eds. Seddon, G. and Davis, M.) Australian UNESCO Committee for Man and the Biosphere, Publication No.2. Aust. Govt. Publishing Service, Canberra. pp. 49—62.

Frankel, O.H. (1976) The IRRI phytotron: science in the service of human welfare. In 'Proceedings of the Symposium on Climate and Rice'. International Rice Research Institute, Los Banos, Philippines. pp. 3–9.

Frankel, O.H. (1977) Genetic resources. *Ann. N.Y. Acad. Sci.* 287, 332–344.

Frankel, O.H. (1977) Genetic resources as the backbone of plant protection. In 'Induced Mutations Against Plant Diseases'. Proceedings of a Symposium, Vienna, 1977. International Atomic Energy Agency, Vienna. pp. 3–12.

Frankel, O.H. (1977) Natural variation and its conservation. In 'Genetic Diversity in Plants'. (Eds. Muhammed, A., Aksel, R. and von Borstel, R.C.) Plenum Press, New York. pp. 21–44.

Frankel, O.H. (1978) Biosphere reserves: the philosophy of conservation. In 'Conservation and Agriculture'. (Ed. Hawkes, J.G.) Duckworth, London. pp. 101–106.

Frankel, O.H. (1978) Conservation of crop genetic resources and their wild relatives: an overview. In 'Conservation and Agriculture'. (Ed. Hawkes, J.G.) Duckworth, London. pp. 123–149.

Frankel, O.H. (1978) Natural resources and technology: evaluation, use and conservation of biological resources. Proc. 3rd Int. Congr., Pacific Sci. Assoc., Bali, Indonesia, July 1977. pp. 303–323.

Frankel, O.H. (1978) Value of wilderness to science. In 'Australia's Wilderness: Conservation Progress and Plans'. (Ed. Mosley, G.) Proc. 1st Natl Wilderness Conf., Canberra, 1977. Australian Conservation Foundation, Canberra. pp. 101–105.

Frankel, O.H. (1978) Philosophy and strategy of genetic conservation in plants. In 'Third World Consultation on Forest Tree Breeding, Canberra, 1977'. Documents. Vol. 1. CSIRO, Canberra. pp. 2–11.

Frankel, O.H. (1978) Germplasm 'preservation'. Plant Genetic Resources Newsletter No. 34. IBPGR–FAO, Rome. pp. 18–19.

Frankel, O.H. (1980) Our evolutionary responsibility. *UNESCO Courier*, May 1980, 25–27.

Frankel, O.H. and Soulé, M.E. (1981) 'Conservation and Evolution'. Cambridge University Press, Cambridge (in press).